"I AM THE SMARTEST MAN I KNOW"

A Nobel Prize Winner's Difficult Journey

"I AM THE SMARTEST MAN I KNOW"

A Nobel Prize Winner's Difficult Journey

Ivar Giaever

Applied BioPhysics, Inc., USA

NEW JERSEY · LONDON · SINGAPORE · BEIJING · SHANGHAI · HONG KONG · TAIPEI · CHENNAI ·

World Scientific Publishing Co. Pte. Ltd.
5 Toh Tuck Link, Singapore 596224
USA office: 27 Warren Street, Suite 401-402, Hackensack, NJ 07601
UK office: 57 Shelton Street, Covent Garden, London WC2H 9HE

Library of Congress Cataloging-in-Publication Data
Names: Giaever, Ivar, author.
Title: "I am the smartest man I know" : a Nobel laureate's difficult journey / Ivar Giaever (Applied BioPhysics, Inc., USA).
Description: Singapore ; Hackensack, NJ : Distributed by World Scientific Publishing Co. Pte. Ltd., [2016]
Identifiers: LCCN 2016024821| ISBN 9789813109179 (hardcover ; alk. paper) | ISBN 9813109173 (hardcover ; alk. paper) | ISBN 9789813109186 (pbk. ; alk. paper) | ISBN 9813109181 (pbk. ; alk. paper)
Subjects: LCSH: Giaever, Ivar. | Physicists--Biography. | Nobel Prize winners--Biography.
Classification: LCC QC16.G49 A3 2016 | DDC 530.092 [B] --dc23
LC record available at https://lccn.loc.gov/2016024821

British Library Cataloguing-in-Publication Data
A catalogue record for this book is available from the British Library.

Copyright © 2017 by World Scientific Publishing Co. Pte. Ltd.

All rights reserved. This book, or parts thereof, may not be reproduced in any form or by any means, electronic or mechanical, including photocopying, recording or any information storage and retrieval system now known or to be invented, without written permission from the publisher.

For photocopying of material in this volume, please pay a copying fee through the Copyright Clearance Center, Inc., 222 Rosewood Drive, Danvers, MA 01923, USA. In this case permission to photocopy is not required from the publisher.

Typeset by Stallion Press
Email: enquiries@stallionpress.com

Contents

Chapter 1	Introduction	1
Chapter 2	Growing up in Norway	5
Chapter 3	The German Occupation	25
Chapter 4	Attending University (NTH)	33
Chapter 5	Adulthood in Norway	53
Chapter 6	Emigration to Canada	63
Chapter 7	Emigration to the USA	81
Chapter 8	Norwegian Vacation	91
Chapter 9	GE Research Laboratory	95
Chapter 10	Friends at Laboratory	109
Chapter 11	My Trip to Moscow	125
Chapter 12	More Research	131
Chapter 13	Sabbatical Leave in England	135

Chapter 14	Back in the USA	155
Chapter 15	Receiving the Nobel Prize	159
Chapter 16	My First Trip to China	179
Chapter 17	Working in Biology	189
Chapter 18	Coolidge Fellowship	195
Chapter 19	Back at GE	205
Chapter 20	University of Oslo	219
Chapter 21	Applied Biophysics	227
Chapter 22	Korea	233
Chapter 23	Lindau	241
Chapter 24	Sports as an Adult	249
Chapter 25	In Business	257
Chapter 26	Today	267
Epilogue		269

Introduction

CHAPTER 1

Life is not fair, and I, for one, am happy about that. That I should receive the Nobel Prize in Physics coming from a tiny village in Norway was very unlikely, and I shall try to describe in this small book, how and perhaps why it happened to me. I do not write a diary and thus must rely on my memory, so the story may be somewhat choppy and not exactly correct, but I promise to do my best. I should say here that I have a brother, John, who is only about a year older than me; we grew up together and we shared most of our memories for the first 16 years or so. But I left for Canada when I was 25 years old, while he stayed in Norway. Sometimes now when we get together to swap stories from our youth, I am very surprised how bad my brother's memory is! So keep that in mind as we go along.

When I think of how things worked out for me, I am amazed at how life depends on a myriad of small things and tiny decisions. For example, if you had left only one second later, you would not have had the accident; if you had not stopped to have ice cream, you would never have met your wife; if your grandfather's first wife had not died, your father would not have been born, and so on. Luck is very important in your life, but I am not the only Norwegian with luck. Mette-Marit is now a Norwegian princess destined to be the Queen of Norway. Before she married the Crown Prince, she had several affairs and a child out of wedlock with a man arrested for drugs. This is not a normal past for a future queen! Life is really chaotic and probably completely random, but most people will not accept that and therefore, perhaps by way of explanation, religion has come into play. In an attempt to come to terms with life's unpredictability, people often think, "It must be God's will, or God's ways are mysterious, but He has a hidden purpose." To those people I say that hopefully one of God's purposes was for us not to enjoy all the religious wars that have happened throughout history. As I understand world religion, the Jews, the Muslims, and the Christians all pray to the same God, and thus they put Him in a difficult position, because which religion should He favor? I am not religious today even though when I was a child my mother told me that when it rained, it was because the angels in heaven were crying. And hanging over my bed there were color pictures of baby angels or cherubs with feathers and wings. Perhaps at the time those pictures were made, artists thought that angels were some type of bird? Yet even today, a real Norwegian-born princess claims she can communicate with angels, talks about their feathers, and has turned her beliefs into a profitable business. She has actually established a private school where people who apply and are admitted are promised to come into contact with their own guardian angel. So lots of strange things still go on with religion in Norway. But in this, Norway is probably a lot like every other place. It is unfathomable to me that people actually continue to kill each other in 2016 because of different religious beliefs. And if you have a military uniform on, you are licensed by your government to kill!

When I was a child, every night I was told to say a simple prayer which sounded something like this: "Dear Jesus Christ, thank you for a nice day; good night mama and papa, in Jesus's name, Amen." But despite my bedroom angels and bedtime prayers, I was fortunately not raised in a strictly religious home; neither my mother nor my father was really religious, and therefore I was not indoctrinated. I believe, as Richard Dawkins, that children are not born Jews, Christians, or Muslims, and perhaps the best thing we can do as parents is to avoid the temptation to make them one or the other. When you get really old, sometimes you want to hedge your bets about religion, and this happened to my mother when she was in her nineties. She asked me a few times about the afterlife. I tried to gently reassure her that she did not have to worry because if there really was an afterlife, she had been good and would certainly qualify for heaven. I am now 86 years old, almost as old as my mother was when she started asking me questions about the afterlife. Unlike my mother, I am at peace with life's randomness. I take responsibility for my own actions, even though — or maybe because — I have had more than my share of luck.

CHAPTER 2

GROWING UP IN NORWAY

Like almost everybody on Earth I had two parents who never went to college or even high school. But they were good and intelligent people. My father was a pharmacist. In order to become a pharmacist at that time in Norway, one had to spend a semester at university after 10 years in public schools. My mother had the same education as my father, except she never went to university. Besides helping out at the drugstore a few times when they were short of workers, my mother was home taking care of her family. At the time that was what most women did. Both my parents were interested in reading, and we had a lot of books. Books then were expensive and I can remember that my father bought books from auctions in Copenhagen, Denmark. I do not think he shopped at auctions to save money. Instead, he truly liked the surprise and suspense of receiving and opening the big wooden crates. He always hoped to find a nugget or two inside the crates, but the contents were mostly cheap literature in Danish. It was okay that the books were written in Danish because at the time there was little difference between

Ivar as a baby, 1929

written Danish and Norwegian. That was due to the fact that Norway had been under Danish rule for 400 years, until 1814. Norway then entered into a union with Sweden. It was about this time that the average Norwegian became literate. Since we had been ruled by Denmark for such a long time, the official written language remained Danish. I actually learned to read the year I had my tonsils removed. I must have had lots of colds when I was young, or the decision to remove my tonsils would not have been made. It was a medical fad at the time, because tonsillectomies are seldom performed today. Today we seem to trust nature a bit more; if you were born with tonsils, maybe they do something important that we haven't figured out yet. At the time we lived in a little village called Lena, but the operation was done is a city called Hamar where they had a hospital. The city was a two-hour boat ride across Mjøsa, the name of the biggest lake in Norway. I was only about 5 years old, and I clearly remember lying on the operating table. The surgeon asked me to count backwards starting at 100. By the time I got to 94 I was out cold. When I woke up, my head was on a rubber pillow and blood was coming out of my mouth. The surgeon and my mother were there, and I cried like a stuck pig. The doctor said, "If you don't stop crying, your mother must leave." So of course I cried even more, and my mother left. The next thing I remember is waking up in a room in the hospital with 4 or 5 adults. I guess there was no children's ward in the hospital at that time. My mother had left me with several children's books. Since I was the only child in that ward, I must have become everyone's little pet; I remember all the adult patients were eager to help me. One book had simple sentences with words missing. You could guess what the words were by gluing the correct pictures into the blanks. After a short time I suddenly recognized I could guess where the pictures in the books should go by myself, and after that I could sort of read.

Some of the first books I ever read were Edgar Rice Burroughs's Tarzan stories in Danish which, for a child about 5–6 years old, was very exciting reading. I just skipped over the complicated names of the apes and other jungle animals, never trying to pronounce them. I simply associated the collection of letters with the characters in the book and didn't get bogged down in name details. Interestingly, to this day I do the same thing, even if the hero has a simple name like John and the heroine is Jane. I can never recall the names when I have finished a book. In Feynman's memoir, if I remember correctly, he wrote about his father who did not tell him the names of birds and animals but just how they acted and behaved and why they did so. So maybe I have a knack for that thanks to the Tarzan books.

Right after I learned to read, my grandfather told me that a great man is born only every 100 years, which made an impression on me. We had a small encyclopedia at home, and so with my grandfather's words in mind, I started to go through it systematically from the beginning, looking for someone great born on April 5, 1829, exactly a hundred years before my birthday. I took my grandfather's word literally as children are apt to do. I could not find anybody, so I decided that I was not to become a great man. But I can't remember if I was disappointed by that new discovery! My grandfather also had a Bible with only illustrations, and the pictures of Noah and his Ark, along with the images of the desperate, almost naked people on top of high mountains facing sure death as the water was rising, fascinated me. Another memory from that picture Bible is Joseph lying at the bottom of the snake pit where his brothers had thrown him. He was apparently unharmed, but horrifically surrounded by vicious-looking snakes and scorpions and his brothers were to blame. This made an impression on me because I had never been able to get along very well with my brother either. He was a year older than me, but I was bigger and stronger than him. So that could not have been easy for him. When we fought, which was rather often, I usually won if it was a fair fight. I actually own that Bible today, but my kids never looked at it because I did not have it when they were small.

Another thing that intrigued me was my father's globe. Much of Africa was colored white. My father told me that the white sections represented undiscovered territory. And he told me the story about Stanley and Livingstone. Stanley went looking for Livingstone in Africa. When he discovered a white man in some place, he reached out his hand and said that famous line: "Dr. Livingstone, I presume?" I thought maybe just like Stanley and Livingstone I could go to some faraway place like Africa when I grew up and discover wonderful, magical things. Or, I could be like Roald Amundsen and Fridtjof Nansen who were great heroes in Norway because of their polar expeditions. Growing up, we were constantly reminded of Roald Amundsen's victory over Robert Scott when they had a "race" to the South Pole. We had been told that Scott used tractors that broke down in the cold, while Amundsen used sled dogs. And that they also ate their dogs in a pinch. I guess the lesson for me was that, although technology could be great, sometimes it is more important to be practical. Also, this was long before Norway found oil in the North Sea. Norway at the time was a very poor and basically an undeveloped country. There was not much we could brag about and so beating the English to the South Pole made us proud.

My parents were rather strict by today's standards. At that time, the adults were adults, and the children were children, and not their parents' best friends. If you have read a book by Robert Paul Smith called *"Where Did You Go? Out. What Did You Do? Nothing"*, you already have a good description of my youth. At the time it was like two different worlds; the adults, lived in one, and the children in the other. As a child you respected the adults, but you did not confide in them. And probably the adults were too busy making a living to try to understand their children. Should the adult world not approve of something that was going on in the children's world, corporal punishment was routinely administered. When my brother John and I were very young, my father sometimes had to slap us on our behinds when he came home from work because of something we had done wrong. Spanking was never done in a rage; we accepted it as much as our parents did. That was how things were done, and there were no hard feelings on either side as far as I can remember. I actually treated my own kids the same way my parents treated me, and today I am very proud of the way they all turned out! But times have changed and I am not sure I would have spanked my children if I were raising them in today's world because it is probably illegal? Looking now at some of my children's and their peers' difficulties with their kids — young people who were not raised with these kind of rules — I am convinced that my parents did the right thing. After all, when you do something wrong, and children always try to test their limits, you have to take the penalty. In life and in science, it helps to know the rules so you can figure out when to obey them and when to try and break them.

When I was about 5 years old we moved from an apartment in a village to a small house on a farm — a rather big farm by Norwegian standards, since it was almost 100 acres. It is therefore easy for me to remember my life before age 5 and after age 5 because of that move. The farm my family moved to had 7 horses, about 50 cows that, by the way, all had names, maybe 10 pigs, and an untold number of chickens. One of my jobs was to go down to the barn every evening to fetch milk in a 3-liter bucket. The milk was stored temporarily in a large wooden crate in the barn, and the milkmaid always plunged her naked arm into the milk to fill the bucket. Today this would be unthinkable. My mother would store the milk overnight and skim the cream off it in the morning.

Before I was 5 years old, I remember that my brother John and I shared a room with a maid. My parents were not rich, but almost every family had a maid then, because that was the kind of work that young girls could get. One maid that I remember was named Sofie and she played the guitar and sang every evening. She was part of a religious sect: the Pentecostal Movement. I later found out that she actually became a missionary and died in Africa. When I was very young we lived by a tiny stream, and to get down to it you had to go down a rather steep hill. The hill was overgrown with weeds, and one particular weed burned your skin if you touched its leaves. So we were not allowed to go down there. My brother John and I had a boy playmate whose name was Tor. One day, having nothing better to do, Tor and John pushed me down that hill, and the weeds seriously burned me. My mother rescued me and was mad at my brother and told him that my father, when he got home from work, would penalize him. Then she gave me a small bag containing five pieces of candy. She specifically told me to eat all the candies myself and not to share any of them with John. As I remember it, I ended up giving him two pieces anyway, in spite of my mother's wishes. At the time, my older brother was my hero, and besides, he promised never to push me down the hill again. I kept this secret from my mother, because I did not think she would approve, and she might also punish John.

It may seem like I was a very rational kid, but I was really terrified of the dark when I was young, and I imagined all sorts of terror while lying in my bed at night. On the wall of that room hung a famous picture; I believe it was from Russia. In the picture, a troika of horses is running while pulling a sled with several people on board. A dozen wolves are chasing the sled in hot pursuit. The people seem doomed, but there is a man in the sled lifting a baby over his head, ready to throw it to the wolves, in order to save the rest of the people. Absolutely not the sort of picture to have hanging in a room where children sleep! This wasn't the only frightful picture from my childhood. Apparently, the dentist we went to came from the same school of child psychology as my parents; on the walls of his waiting room hung gruesome pictures on how to extract teeth with enormous tongs while friends held you down. As far as I can remember, even though the dentist was my father's friend, my father never went to the dentist. Who in their right mind would have that horrid imagery on display? This illustrates the two worlds very clearly; the adult world was not very sensitive to children. The gap between our world and my parents can be seen in their attitude towards food as well. John and I ate breakfast and evening meals in the kitchen with the maid, but dinner was always a formal meal that we ate in a dining room with our parents. As a child I had to eat everything served, and clean off my plate. If we did not eat the food, it would be served again for breakfast the next day. I remember a soup made from goose that was very fatty that I struggled with several times at breakfast. I liked dessert and could not understand why my father did not care much for dessert, but that he loved soup. Now that I am an adult, I am the same way. The only food I did not have to eat as a child was a particular Norwegian fish dish, "Lutefisk", because my father did not like it. It was only served when my father was out of town. Today that dish is one of my favorite meals. So I inherited my love for soup from my father and the Lutefisk dish from my mother. Maybe it has something to do with DNA?

Like most children, I was very excited about my birthday not only because of presents, but also because I could decide what we should have for dinner; I always chose pork patties. The present on my third birthday also stuck in my mind: a blue tricycle. I became very excited, and wanted to try it outside. But a lot of snow had fallen and my mother said no. So I went outside without my tricycle and made a snowball, as the snow was rather wet. Then I put the ball on the ground and began rolling it forward. To my surprise it became bigger and bigger, and some dead maple leaves stuck to the ball. At the time I remember feeling proud and excited about my accidental invention. When I went back inside to report my findings, I was told that people build snowmen by rolling snowballs. But I was thrilled by my discovery nevertheless, because I had done it on my own, nobody had instructed me.

We had a simple outhouse where I lived the first 5 years with two seats. On the farm there were two outhouses, one with 4 seats and one with 6 seats. In each outhouse one or two seats were built low to the ground and thus were suitable for children. Going to the toilet was often a community affair in Norway in those days, and the thinking was why not, since we also eat together! A common sight inside outhouses was a picture of the Norwegian king, King Haakon VII, clipped out from a newspaper. And speaking of newspapers, old newspapers were of course used for toilet paper. You were also guaranteed to see some dirty sexual words written on the walls — words that we never were allowed to say aloud. At the school we had a row of outhouses. The first three outhouses were for boys and the next three for girls. One of my friends used to actually stick his head down into the hole on number 3 so that he could peer over at the bottoms of girls sitting on toilet number 4. I never could do that; I was curious of course, but I could not stand the smell! After we moved to the farm, my family had an indoor toilet also, which was rare at that time. The house we were renting had been recently built by the farmer himself, who would move in when he got too old to work the farm, and he knew the ropes. The farm would then be given to his oldest son; a daughter could only inherit the farm if there were no sons. This was true for the Kingdom of Norway as well, but it has recently been changed to the oldest child. We are slowly moving into a fair world.

As a child I always had a technical bent. For example, I loved to take apart old clocks, locks, and farm machinery just to see how things worked. We had a record player that you had to crank up in order to play a record. It was a wonderful piece of equipment. I got a lot of satisfaction when I figured out how the cranked-up record player controlled the speed of the record. There were three weights attached to levers on a rotating, vertical shaft. At standstill the weights hung straight down. But when you started the record player and the speed picked up, the centrifugal force overcame the gravity, and the weights would rise and move a lever. The lever then applied a force to the brakes that slowed down the rotation. This made the weights start to go down again, but then since there was no braking, the rotation sped up, and so on. Wonderful! In a way I feel sorry for kids growing up today, because they don't have the opportunity that I did to figure out how things work. For example, if you take apart a modern digital watch, you have no chance of figuring out how it works, let alone repair it. If you open it up, all you would see is a battery and one or two integrated electronic chips. When I grew up, you would find a wind-up spring that provided the energy to the tiny gears that moved the visors. The speed of the wheels would be controlled by a pendulum for a wall clock, or a wheel rocking back and forth attached to a soft spring for a regular clock. Because of my tendency to take things apart, and to fix broken things, my mother thought I should become a locksmith.

Public schools

When Norway separated from Denmark in 1814 they introduced seven years of obligatory schooling. The schools were good for that time.

Unfortunately when I started school in 1935, nothing much had changed inside the classroom, though a lot had changed outside. The curriculum was more or less the same as it had been more than 100 years earlier. The school I went to had only three teachers, and they taught all subjects. One teacher was responsible for teaching the first three grades, and thus each grade could only meet twice a week, since there were no mixed-age classes. Saturday was a regular school day. There were two teachers for the next four grades, so those grades could meet three times a week.

Norway then and now has a state religion — Lutheran. For the seven years that I went to public school, religion was taught the first hour of the day, and thus was considered the most important subject. We basically learned the New Testament. I remember I tried to figure out how Jesus could walk on water; I decided it must have been on ice, which of course is common in Norway. I have since learned that the coldest temperature measured in Israel in modern times is $-12°C$, and so given that fact, maybe there was a short cold spell during Jesus's lifetime and he really did walk on water! It was more difficult for me as a child to understand how Jesus fed 5,000 hungry people in the desert with only five loaves of bread and two small fishes. When I asked my father about this, he explained it by saying everybody probably brought a little bit of something.

Even as a child I never believed outright what I was told, I had to figure out if it made sense compared to what I had learned before, similar to figuring out how to make a big snowball. This did not sit well with some adults, but both my father and mother accepted it well. Religion gave me the most problems as for example, The First Commandment: *"You shall have no other gods before me."* I found this very puzzling, because it seemed to me that there were at least three gods: God, Jesus, and the Holy Spirit. And then there was Satan, who tempted Jesus and who also seemed like a godlike figure. It didn't help that the teacher did not allow much discussion. We also had to learn psalms by heart, and they were written in old-fashioned Norwegian language (which was really Danish) and difficult to understand. In addition to religion we had to learn penmanship, arithmetic, and Norwegian. Sometimes we also were allowed to draw in class. We all drew small houses with a fenced-in garden and a flagpole with a Norwegian flag on top. But one day I decided to draw a tractor with big red rear wheels. The teacher praised me, and then the whole class changed to drawing tractors!

Sometimes we were asked to read aloud in class. That was a big problem for two of my classmates, because they never learned to read. Since they were not dumb, looking back I think they must have had dyslexia. Of course at that time no one knew anything about learning disabilities. Or maybe they couldn't read just because they couldn't see well. Problems with vision were rarely corrected in those days. I was one of the exceptions. After I was diagnosed as being cross-eyed, I had to wear glasses for one year. At that time in Norway, the only people who wore glasses were very old people if they were lucky enough to afford them. I guess I was lucky, but it did not feel that way. Kids who wore glasses were considered freaks; I was bullied and called bad names like "four eyes" by other children. It was very common at that time to tease or bully people who were different. I remember one unfortunate boy who was incontinent and was called "Pisspotty". When I think back I tended to defend the people that were made fun of, and that probably came from the fact that I was teased for wearing glasses. When Norway was occupied by Germany in 1940, a big surprise to me was that so many young soldiers wore glasses!

We did not have much homework, and the standard excuse for not completing it was "our light bulb burned out" instead of the famous American excuse: "the dog ate my homework". In fact, however, a few homes in my neighborhood did not even have electric lights when I was growing up. We always skied to school in the winter. We had an unwritten rule: the person who lived farthest away from the school stopped by the house of the person who lived the next farthest away and picked up the person who lived there, and so on. In this way our parents could feel reasonably sure that we got to

Graduation from public school, 7th grade, 1942

school unharmed. It was a simple solution compared to today where every parent seems to drive their children everywhere. When we left in the mornings in winter, it was often pitch dark. There was lots of snow, and it was very cold. The coldest I can remember was in 1944 when it was −44°C. That was the time when the German offensive collapsed in the Soviet Union. It was reported that they used dead, frozen-stiff Soviet soldiers as road signs. When we got to school we would examine each other. If you had frostbite, which would appear as a white spot on your face, we would rub it warm using snow! It was common knowledge then that the best way to treat frostbite was to apply heat gradually. But I have since wondered if it would not have been better to just skip the snow step and jump right to rubbing the spot with a mitten. Maybe I can look it up on Wikipedia someday?

Work on the farm

Every fall there was a three-week "potato vacation". Of course this was not a vacation in any sense of the word, and today it would be called child labor! For three weeks classes were cancelled and all the kids had to go out and pick potatoes instead. The farmers had a machine that had a spinning wheel with spokes, which was pulled by two or three horses. The first job I ever had was to ride the front horse, making sure that it helped pull the machine. The object of the spinning spokes was to dig the potatoes up from the ground. The potatoes were then flung onto the field. Children and some older women then set about picking up the strewn potatoes. The length of the real estate you covered determined how much you were paid. It was really hard work, for as soon as you had picked up the potatoes nearby, the horses and the machine were at it again, spraying more potatoes all over.

Another job my brother John and I often had in the fall was working as "in between" drivers. Every fall adult workers would load hay onto a wagon hitched to a horse. Then a boy would drive the horse and wagon to the barn, where adult workers exchanged the loaded wagon for an empty one that needed to be driven back to the field. In the spring, instead of hay, cow manure was loaded onto a sled and we drove it on the snow from the barn to the field. Adults in the field unloaded it to fertilize the field. Or maybe, since it turns out cow manure is not a good fertilizer, they just wanted to get rid of all the manure that had piled up under the barn over the winter! You had to stand on the sled, keep your balance, and try not to fall into the manure. But that was bound to happen a few times.

Of course, such types of work involved harnessing the horses. This was difficult because farm horses are really mean. You see, when a boy tries to harness a horse, the horse knows the boy is small, and strongly objects to being harnessed. Sometimes the horse would try to both bite you and kick you. Meanwhile, while you were struggling with the horse, you could count on the real farmhands to be standing around and laughing at your efforts. Sometimes in the summer we went to fetch the horses in the morning, because they were put out to pasture at night. We then rode them bareback back to the farm. Once the horse my friend was riding went straight for the stable, and just like in cartoons he hit the wall over the door. But, unlike in cartoons, he was seriously hurt.

I used to save most of the money I made, but sometimes I went to the store to buy licorice which is still my favorite candy. The licorice cost roughly a penny for one small rod. I made my purchases in a small grocery store. The owner stood behind a counter and went to personally fetch whatever the customer wanted. If you bought sugar, for example, he first had to weigh it out for you, and then pack it. When I bought just one stick of licorice, he had to open a trap door in the floor, walk down a set of stairs, pick up the licorice, carry it upstairs, and give it to me; so much effort, all for only a penny! I grew up in a very different world compared to the world today. Although Norway back then was a country with a central government, we had no sense of that. In our community we had a tailor, a shoemaker, a cabinetmaker, a blacksmith, a doctor, etc. We had all we needed, and our little place managed to get by with very little interaction with the government and the outside world. In a sense you could say it was more of a medieval than a modern place.

One year before Christmas we visited a small city, Gjøvik, and I saw a movie for the first time. The movie was Walt Disney's "The Three Little Pigs" and it made a big impression on me. Since I was scared of the wolf, I decided then and there that I would live in a brick house when I grew up; as a matter of fact, my house in Schenectady, New York is made of brick. Also on the same trip I saw a coconut for the first time, and all I wanted for Christmas that year was a coconut. My parents bought me one and packed it in a square box. My father promised me a 5-crown piece if I could guess what it was. By shaking the box I could hear the liquid in the coconut, but I did not know that a coconut had liquid in it, so I was unable to figure it out. We always celebrated Christmas at home with my father's parents who lived close by. And at Christmas dinner my father served aquavit which is Scandinavian schnapps. My grandfather was limited to two glasses, and he enjoyed those very much. Christmas was a big holiday for me when I was very young. Before Christmas we made decorations for the tree, and we used special color papers to weave what we called Christmas Baskets to hang on the tree. My father always filled them up with whatever candies were available. We also made colored balls, by carefully cracking walnuts into two, eating the nut and gluing the halves of the shells together, attaching a string and finally painting them silver. The presents we gave our parents were always homemade, mostly made by sawing various shapes with a copying saw. One tradition my father enjoyed was to dress up as Santa Claus and give presents to the "nice kids". When we got older we recognized him of course, but the tradition continued. I have since taken that job in my family, except I always let the children recognize me, as I will not deceive them with supernatural things. We also celebrated Easter, and then if we put our caps outside, the Easter bunny would always lay marzipan eggs in it. When I was 6 or 7 I decided that the Easter bunny did not exist, and I refused to put my cap outside. John, my brother, was smarter and did put his cap out. He got richly rewarded with Easter eggs, and I got none. Next year I decided to believe in the Easter bunny again!

Normally after the first snowfall the farmer hired help to slaughter a few pigs. A couple of long wooden tables were set up in the snow to hold the carcasses of the pigs. Before the pigs were killed, masks were placed on their heads. I believe that once a mask was put onto a pig's head, a big hammer was used to drive a stake attached to the mask into the pig's brain. This operation was always done behind the barn, so (perhaps thankfully) I never actually witnessed it. But even those of us not present for the killing could always hear the pigs squeal for a minute or two prior to their death. The carcasses were then laid out on the wooden tables. As the bodies bled, the fresh new snow was colored with red spots. Most of the blood was, however, collected for later use, such as making blood pudding. Lots of hot water was then used to remove the hair from the pigs' skin. Pigs were used to make sausages, canned meatballs and gammon. Gammon is the hind leg of the pig that is put into a strong salt solution, then rubbed with salt and hung for maybe a whole year to cure before it is eaten.

Another service performed on the farm was the fertilization of young heifers. There were about 50 cows in the barn and one very big bull. In the summertime when the bull was let outside to graze, we never dared to go into the enclosure, because the bull was known to be mean. I can understand why the Spanish people get excited about bullfights, because the bull's behavior is really unpredictable. People would come with one or two heifers to the farm to get them impregnated. There was a stall outside the barn where they placed the heifer. Next, a couple of farm workers went into the barn and fetched the bull, holding him with a big rope that was tied around his neck. As soon as the bull saw the heifer, he would be eager to mate; the bull would jump on the heifer from behind. I was always afraid that heifer would collapse under the weight of the bull, but the whole affair was over in about 15 seconds. This was my first introduction to sex. I learned much later on that fortunately, people use more time when they procreate!

Sports

When I grew up sports were important for children, but not so much for adults. During wintertime it was all about cross-country skiing and ski jumping; during summertime it was all about football (soccer in the United States). The children themselves mostly organized these sports since adults had little interest in, or maybe time for getting involved. Close to the center of the village there was an empty field, which we used for playing soccer. After school a few boys showed up, and as soon as the guy who owned the actual football came, teams were selected. The two oldest boys chose teams with the boys who were there at the time, and then each team absorbed other boys in the order they arrived. We used a couple of stones to mark the goals. There could be any number of players on each team and our rules were different from those used in a classic soccer game.

For example, "offside" was rarely called and instead of corner kicks, three corner kicks were saved up and then exchanged for a penalty kick. If the game had gone on for a long time, we sometimes chose new teams. The two best players then chose the players alternately, and of course this was a time of stress, because everyone wanted to be chosen early and not last. But alas, there were no psychologist present on our field and therefore some boys were always unhappy but tried not to show it.

As for skiing, cross-country ski competitions were never arranged for children and I believe you had to be 18 years old to participate. For some reason in ski jumping there was no age restriction. Almost every boy ski jumped in the district where I lived, and our hero was a Norwegian ski jumper named Birger Ruud. He had won Olympic gold medals in both the 1932 and 1936 games. I was fortunate compared to other kids, because I often got jumping skis and bindings from my parents for Christmas. We lived maybe a couple of miles away from a rather big ski jumping hill where, I believe, you could jump maybe 150 feet or so. The hill had a wooden tower, and it was not easy to carry your skis to the top, because the wooden steps going up were covered with ice and snow. Contrary to popular belief, ski jumping is really a very safe sport, as long as you organize it properly. We kids took care of the organization ourselves. Since the people on the top of the hill could not see if the landing area of the hill was clear, we always had a person to signal that it was clear before anyone jumped. In all my years of ski jumping, I only witnessed one accident — a very good ski jumper used only one ski on purpose and landed on the wrong foot! If you look at professional skiers, at the end of a season maybe only half of the people participating in downhill skiing are left with no injury. Meanwhile, almost all the ski jumpers are fine and still competing.

Once each winter three different neighboring public schools arranged a ski-jumping competition for the boys in 4th grade or higher. Girls never participated — they only watched. Every boy was compelled to participate; otherwise his school would lose points. There were always a few boys who were afraid, or who did not have the proper skis, clothes, or boots. I am ashamed to recall today that after lending them whatever they needed, I, along with a few others, would forcefully push those boys down the ski-jumping hill. I always won third prize every time I participated. In ski jumping, your jump is judged both by style and the length of the jump. On this hill you could jump maybe at the most 100 feet. It was an exciting event, and in preparation for it almost every boy made his own secret ski wax. I remember I melted old broken gramophone records and then mixed them in with some melted candles. When I was 16 years old I ski jumped on the biggest hill in Norway at the time, and cleared 216 feet. Back then the best ski jumpers could jump 270 feet. Today, of course, in ski flying, people land at over 800 feet away. Ski jumping was my favorite sport, but in 1943 during the German occupation, the Nazis started organizing the sport. Patriotic Norwegians then destroyed all the jumping hills in protest, mostly by cutting or sawing down the wooden towers. For this reason I did not jump for about 2 years. When I started again after the war, I had become nervous and afraid. Then you don't do very well, and I stopped competing even though I really loved to compete. In general people do not know that science is very competitive, and therefore stressful, but for the competitive person, science is also exciting and fun.

Mythical Creatures

As a child I also liked to walk in forests, and there were a lot of forests around where I grew up. My father was an eager mushroom picker and taught us how to recognize the edible mushrooms, and keep away from the poisonous ones. Sometimes I stayed out in the forest overnight with a sleeping bag. I remember one time I was sleeping on a small path in the forest when I was awoken by a thundering sound. It turned out to be five or six horses running on the path. In Norway at that time, horses were allowed to run freely in the forest during the summer, because there was no work for them to do on the farms. People in Norway believed that horses never stepped on a living creature while running, but I was not brave enough to test that theory! That night I instinctively rolled out of the way and thankfully was not harmed. There was always one stallion in the small pack of mares. Many people would come to watch the spectacle when the horses were let out for the first time in the summer. The freed horses attracted a crowd because people wanted to watch the stallion have intercourse with the mares. If you think that is peculiar, this spectacle is often broadcast today on the TV news in Norway! The farm culture from the past still runs deep in modern Norway; you could argue that we are not very sophisticated.

I sometimes brought a fishing rod but I never caught many fish. I loved to sit by a lake watching for "nokken". "Nokken" is a mythical, Norwegian being with green eyes and is believed to live in small, bottomless lakes. A Norwegian painter named Theodor Kittelsen made him famous. "Nokken" probably originated from someone viewing a rotten fluorescent log dipping in a lake. We have lots of mythic beings in Norway. For example, trolls are widely known outside Norway. Abroad the trolls are often portrayed as cute little figures, but when I grew up the trolls were very big, mean, and often had only one eye shared between several trolls. Since they could not stand the sun, the best way to trap a troll was to delay it such that the early mornings' sunrays hit it; then the troll would turn to stone. "Hulder" is another example of a magical creature, which is not as well-known as trolls. "Hulders" are seductive maidens who also live in the forest. My grandmother claimed she saw a "hulder" once. They are beautiful girls, except that they all have the tail of a cow, which they sometimes manage to hide from the observer. According to legend, in order to escape their hold on you, you must throw a knife over their head! We also had "nissen" who was supposed to live in the barn and help the farmer take care of the animals.

On the farm where I lived, the farmer always put a plate with porridge at the barn entrance on Christmas Eve for "nissen". In modern times "nissen" has morphed into "Julenissen" which is equivalent to Santa Claus. Since I grew up in the farm country in Norway, I was not affected much by "draugen", who is a scary creature living in the oceans and is the major cause of fishermen's shipwrecks and drowning. It was not easy to grow up between all these scary beings and in addition we also had regular ghosts; there was no way I would have approached a graveyard at midnight. I can no longer remember when I stopped being afraid of the dark; I may have been a teenager. And I know I believed in Jesus longer than I did Santa Claus, but I might have been about 10 years old. I know for a fact that I was confirmed in the Lutheran church when I was 13 years old, and the minister asked me a lot of questions in the church. He asked mostly John and me because he knew we would know the answers. I was tempted at the time to point out how ridiculous religion was, but I did not want to embarrass my parents. My sister Mette, who is 9 years younger than me, however refused to be confirmed. She was brave that way, and still is.

THE GERMAN OCCUPATION

CHAPTER 3

Norway was occupied by Germany for five years. Germany attacked Norway on April 9, 1940, to everybody's surprise. The German excuse was that they wanted to protect us from the English. But since the Norwegian Queen was born in England, we had good relations with the English and did not need to be protected. Norway was not prepared and thus was an easy target. Prior to the occupation, my mother used to save articles from the newspapers about the brutal winter war between the Soviet Union and Finland in 1939; she strongly felt that this was as close to war as Norway would ever get. Little did she know…

I was 11 years old when the war started and 16 when it ended. It is the only occupation I have experienced firsthand. Fortunately, the Germans behaved very well in Norway, perhaps because we Norwegians were more Aryan-like than they were. For example, in the village where I lived there were a few restaurants, and each time a German soldier entered or left, he always saluted at the door. They did this by turning around, clicking their heels together and raising their right arm in a Nazi fashion. Meanwhile, we boys laughed at their formality. When they first arrived, the German soldiers bought chocolates and smeared butter all over it before they ate them; we thought they were crazy. This made such a big impression on us that by the end of the war we were all still talking about it, only now with envy. Now people would say, "Remember what the German did with the chocolates? That must have tasted wonderful!" There was not much to eat towards the end of the war, but at least our family never starved. I remember eating porridge made with water rather than milk in the evening, and meals with meat and fish were rare. Sometimes we got horse meat from slaughtered workhorses on a farm. The horse meat came in small pieces and was very tough to chew. But despite the meat shortage, thanks to living on a farm, we always had potatoes, cabbage, carrots and rutabaga. Since rutabaga — not that familiar in the US — is very rich in Vitamin C, we referred to them as Norwegian oranges.

The farmer from whom we rented our house became a Nazi during the war, along with 10% of other Norwegians at the time. Probably another 10% of the population became active in the so-called "home-front". The home-front was a secret organization of patriotic Norwegians that actively sabotaged the Germans during the occupation. The Germans issued a directive that if any German soldier was killed, they would retaliate and kill 10 local people. They even named the people they would kill, and it was usually prominent people in the local area like the doctor, the priest, the teacher, etc. They seldom carried out their threat, but they arrested the people and put them in concentration camps. Despite this serious danger about 90% of Norwegians were against the Germans. The Germans themselves thought all of us were their friends. Right after the war I hitchhiked through Germany and got a ride with a truck driver. We spoke a little; he had spent most of the war in Bergen, Norway. When he found out that I was Norwegian he said, "Too bad we lost the war." I said, "Yes," because of the position I was in. I could not explain to him what most Norwegians including myself felt about Germany at the time. I was afraid he would throw me out of the truck. In my opinion, the problem with occupations of any kind, at any time in history, is simply that people don't like to be occupied. No matter the circumstances, the occupying forces will not be liked. So even though many Americans may believe that when they went to Vietnam, Iraq, or Afghanistan it was in friendship, I am sure, based on my own experience, that the majority of the people in those occupied countries did not and do not believe that. Good intentions do not count in such times. Remember, the Germans told us Norwegians that they occupied us to save us from the English!

During the war the Germans enforced a complete blackout. This meant that everybody had to have heavy black paper curtains in front of all their windows. As a result, both the Northern Lights and the stars were very visible when you looked at the sky while travelling to and from school or later at night. There was a big rumor about the Black Lady who many people claimed to have seen. She was supposed to wear black skirts and was a very fast cross-country skier impossible to keep up with; some kind of ghost really. I never heard of her hurting anybody, but we children were very afraid of meeting her.

Meeting Inger

I started middle school during the war when I was 13 years old. When I entered my second year of middle school, a new crop of students started and Inger was one of them. I was sitting with some friends in a small, pyramid-like shelter that was built over a well. It was a good place to peek through cracks in the structure at the new girls entering the school while not being seen by them. They were running on the field unaware that they were being observed. There was a beautiful long-legged girl from a small village called Skreia who was very gracious. I pointed her out to my friends, and declared that I would date that girl. I kissed my first girl at the end of seventh grade. Inger ended up being the third girl I kissed. When one is young, one

Inger and Ivar 15 years old, 1944

falls passionately in love. The first time I dated Inger, we went for a walk in her village. She was to be home by 9 P.M. We did not have a watch and made it back half an hour late. Her father was waiting for her at the door. Since she was late her father slapped her on the cheek so that I could hear it. Inger told me later that she was very embarrassed by this, but it was something I was used to from home. Despite the drama of young love and living mostly apart in our youth, Inger and I have more or less been a pair since that time long ago when we were both 14 years old. For example I spent 4 years at the university while Inger spent some time as an au pair girl in England. We wrote letters and sometimes met in the summer. Strange as it may sound, we are still in love at age 86, and I am not bragging.

Secondary schools

Normally in Norway at that time, you went to middle school for three years before starting high school or gymnasium as we called it, which lasted another three years. But a new rule made it possible to skip the third year of middle school. Since my grades were good, I was able to choose that option; this meant that I could start gymnasium at the same time as my brother John who was one year older than me. Education for most young people didn't last long in those days. Out of about 20 children in my grade in public school, only three went on to middle school. After all, by the time you started middle school you were 14 years old and considered an adult. Most of these "children" ended up becoming farm workers. This pattern was repeated in the transition from middle school to high school; again we were about 20 students total, and again only three or four of us went on to high school. In other words, only about 1 in 50 children finished high school. Today, Norwegian children are still only required to go to school when they are 6 to 16 years old; high school is still voluntary. Since there was no gymnasium where we lived, my brother and I had to move to a little city called Hamar, which was a 2-hour boat trip across Mjøsa, the same city where my tonsils were removed. We rented a room together. We had breakfast and evening meals in our room and ate dinner midday at a cheap restaurant in the neighborhood.

War ends

By the end of 1944, it had become clear to everyone in Norway that Germany was losing the war, and therefore, Norwegian unrest with the occupation increased. Many of the high school teachers in Hamar were arrested for refusing to sign some kind of German declaration. A couple of people were shot right outside high school, and the school eventually had to close. Since I had gone directly to high school from the second year of middle school and not finished the third year, I went home in Christmas 1944 to do that while John remained in Hamar. We moved out of our room and John moved in with a friend hoping for some private schooling.

The war in Norway ended on May 8, 1945. I remember the moment clearly. I was standing on the road outside the middle school talking with a good friend, a girl named Jorunn, when a truck came along. The truck driver yelled out to us, "The war is over!" I immediately started to cycle home, which was about three miles away. However, the same truck had stopped to refuel, which meant the truck driver had to fill the wood gas generator with more wooden chips from a sack he had brought on his flatbed. During the war there was no gasoline in Norway, so taxis and trucks got outfitted with wood-burning gas generators. A wood gas generator is a gasification unit which converts wood chips into wood gas consisting of nitrogen, carbon monoxide, hydrogen and traces of other gases, which after cooling and filtering, can be used to power an automobile engine. The lack of power prevented these vehicles from moving fast, and so of course there was always the temptation to hang on to taxis or trucks while you were riding your bicycle. Normally the drivers hated it when we kids hung on, since of course despite the slow speed, it was still dangerous. But given the momentous occasion, the end of the war, this time the truck driver encouraged me to grab onto his truck while we both yelled out to everyone and no one that we passed, "THE WAR IS OVER! WE WON! WE WON!"

When I reached home I was met with a myriad of empty cans and bottles piled up in the hallway and on the steps. My mother had thrown out our family's "treasures" from the attic! During the war when we had run out of everything, empty cans and bottles were needed to buy anything. For example, during the war, people could not buy medicine at my grandfather's drugstore unless they brought an empty bottle or a can with them. So of course we saved lots of them in the attic. In an elated mood now that the war was finally over, my mother had dramatically thrown the whole lot of our "treasures" down the stairs.

In school the next day, two of my older classmates came in carrying guns and wearing armbands that indicated that they belonged to the resistance. This signified that they were part of the anti-Nazi home front. I became very envious, because the guns captured the teachers' attention, and of course the two boys carrying the guns got all the attention from the girls. During the war the Germans had collected all the Norwegians' guns, but some had been hidden and not turned in. The Germans had also collected all our radios; this meant that we could not listen to British or other foreign stations. But now that the war had come to an end, a few radios also came forward, and we tried to receive foreign stations on shortwave radio. I remember the only two radio stations listed on the shortwave scale were Hilversum and Schenectady! Who would have predicted then that I should one day be sitting in Schenectady and writing this book? There are many strange coincidences in this world. No wonder people believe in déjà vu and miracles!

After five years of war, Norway was in a strange state for the next several months. At first the American and English soldiers were very popular in Norway. Not only had they freed us from the occupation but they also gave away cigarettes, chocolates and nylon stockings. However, their popularity did not last very long because they were not as disciplined as the Germans. For example, they caused many traffic accidents with their jeeps. They did, however, have a better chance with Norwegian girls than the Germans had, which meant that the Norwegian men were envious. The girls who had been dating the Germans during the war were now punished, mostly by forcibly cutting their hair off. I really felt sorry for them as they were mostly girls from poor families. They tried to hide the damage by wearing a scarf on their head, but since we lived in a small community we of course knew who they were.

There had been no death penalty in Norway since 1902, but that changed as a result of the war. At the beginning of the war, King Haakon VII and the Norwegian government escaped to England. Norway was then "ruled" in absentia. For example, during the war, in 1941 and 1942, the escaped government in England decided to institute the death penalty in Norway for people who had collaborated with the Germans during the war. This was probably not legal, but the newly elected parliament in 1945 made it legal by vote. For a country to make an exception to its long-standing stance against capital punishment reflects just how much Norway hated being occupied by Germany. Looking back, I can't help but question if it was an appropriate response. There is no longer a death penalty in Norway. In fact, the maximum prison sentence is now 21 years, even for Anders Breivik who shot 77 young people on July 22, 2011. Quisling, who was the head of the political party "Nasjonal Samling", similar to a Nazi party, and who "ruled" Norway during the German occupation, was tried and shot, along with 46 other people right after the war. The name 'Quisling' has now become an English word that means a traitor. Nazi collaborators who were tried early on after the war got the heaviest penalties, but after a while collaborators were basically forgiven and forgotten. For a long time though, there were two kinds of people in Norway: the German sympathizers and the rest of us. Since the Swedes in 1940 helped the Germans by letting Germans soldiers pass through Sweden by train so they could fight in the northern part of Norway, the Swedes were not popular after the war. Germans and Swedes are still viewed with some suspicion. To this day the main newspaper in Norway recommends that Norwegians who travel to Denmark carry a visible Norwegian flag; otherwise they could be mistaken for Swedes! The Danes did not like the Swedes either, because the Swedes were for Germany during the first few years of the Second World War when Denmark was occupied by the Germans. I can, however, think of worse fates than being taken for a Swede!

Attending University (NTH)

CHAPTER 4

Preparation for the University

I finished the three full years of middle school in the spring of 1945 and went back to Hamar in the fall to finish the last two years of high school (or gymnasium as we called it). I was a very good student and both mathematics and physics were easy for me, but I had more problems with languages. My parents had instilled in me the idea that I should attend a university, probably because none of them had the opportunity to even go to high school. Despite my parents' lack of formal education, they were highly knowledgeable compared to most of the people in our neighborhood. Like most young people, I had doubts about what to do with my life, so to please my father I applied to become a pharmacist, and was accepted at Oslo University, the only university in Norway at the time. At that time there was great competition to enter most of the programs at the university. I also applied to the Norwegian Institute of Technology which educated all Norwegian engineers. My mother encouraged

Ivar as a high school graduate, 1947

me to try to become an engineer. My first choice was to study electrical engineering, second to become a chemical engineer, and third to become a mechanical engineer. You were only allowed to apply for three choices. I had excellent grades from high school; I had only two Bs, one in French and one in New Norwegian (a secondary Norwegian language). With the two Bs, I only qualified to become a mechanical engineer. I could have become an electrical engineer if I had chosen to go to the University of Copenhagen in Denmark. At that time I honestly believed that NTH (the Norwegian Institute of Technology) was the very best engineering school in the whole world. I was brainwashed like most Norwegians my age. To illustrate this let me tell you about Gro Harlem Brundtland. She is a little

Ivar with with a student cap typical for Norway, 1947

younger than me and became the prime minister of Norway. While giving a speech she said in all seriousness: "It is typical Norwegian to be good!" At the time all of our information came from Norwegian newspapers and a single radio station run by the government. I remember when I saw American films after the war, I was always surprised that whenever somebody turned on a radio it would be playing popular music. In Norway at the time, dance music was only broadcast from 10 pm to 11 pm on Friday and Saturday evenings.

A prerequisite to enter NTH was one year of industrial experience, which I received at the Raufoss Ammunition Factory. This factory was only about one hour away by train from my home. This became a very valuable year for me, and I think it is too bad that it is no longer a requirement for engineering study. First I worked in a tool crib where the workers came to borrow the tools they needed to use next. The guy in the tool crib was very stingy and made it difficult for people to get replacements for whatever tool they needed. He acted as if the tools were his personal property. Later I worked in similar places in both Canada and USA and noticed very similar behavior among people working in the tool cribs. I think that your profession helps to form your personality, maybe even more so than your nationality.

Then I became a toolmaker and learned to operate a lathe. I became very good at it. I loved the work, and tried some small inventions during my lunch hour to automate parts of the lathe. The foreman became impressed by my efforts and started to use me to set up the lathes for other workers. At the time, you had to change the gears by hand to adjust the speed, and I was good at making the right calculations. It was a dangerous place to work, as all the lathes were driven by belts from a wheel up in the ceilings, and it was easy to get in the way of the belts. Since you basically learn how pieces of metals (tools) are made, you later on understand how to design pieces when you become an engineer, such that the tools are easy to make. You also learn to be careful; once I nearly killed myself using a milling machine. My steel milling tool was maybe as thick as a finger, and I was feeding the machine in the wrong direction. I had been warned of that, but it is easy to make mistakes. My tool snapped into two pieces, and because of the elastic recoil, it hit the ceiling with a great force, after having grazed my chin.

After six months I changed jobs and went to the foundry. Now I had to learn how to make forms in sand so that later when molten iron is poured onto the forms, you get a piece of the right shape. We were paid according to how many pieces we managed to make. That was great, because as an apprentice I did not make much money. So I worked very hard the first week and made 17 pieces; I felt very proud of my effort. The next Monday morning three big workers showed up at my workplace and told me, "Nobody, but nobody, makes more than 12 of these pieces in a week, do you understand that?" Well, the guys were very big, so of course I understood it. Unions were and are very strong in Norway. And for the rest of my stay in the foundry I always checked with the people in the know before I did finish something. For reasons that are hard to understand for me, many people like to finish their work early, so they can shoot the breeze, read the papers, or fill out soccer gambling slips, lottery tickets, etc. the rest of the day.

Later I worked where they cut out small metal disks that were later formed into casings for ammunition. There they had big presses, with pistons that came down, hit a metal plate and cut out several circular pieces of metal which were used for the casings. To prevent injuries the engineers had rigged up a contraption such that a shield came down first, so that you had time to remove your hands before the pistons came down. Of course if you disengaged the safety device, you could work about twice as fast, and three of the seven people who worked there had missing thumbs. That was reason enough for me not to attempt that trick! I also had the opportunity to test the ammunition we helped make. It was done by firing from a stabilized gun. If the ammunition hits the target within a small circle the ammunition qualified for NATO countries; if not, it went to Eastern Europe.

When I started working at the factory I lived at home and took a train back and forth. That took a lot of time, so I decided to rent a room in Raufoss when the winter came. I got a nice room on the second floor in a private home. Beside the room was a little cot where I had a washing basin. There was of course no central heating, but I had a wood stove in my room. Many times in the morning the water was frozen in the washing basin. A widow owned the house, and she had a daughter roughly my age, who liked me more than I liked her. It's strange now that I am 86 years old, I think every 20-year-old girl is beautiful, but when I was 20 that was not the case. I was actively pursued by the girl and may have succumbed, but fortunately a high school friend, Bjorn, also worked at the factory. He had obtained a room supplied by the factory, and I escaped her by moving in with him.

We had a good time together. We ate at the factory cafeteria which had ample amounts of simple food. There were no choices; you ate what was served that day. My friend was also headed to NTH for Chemical Engineering so he got his training in the chemical part of the factory. For this reason he had access to 'red spirit', which was denatured alcohol. It tasted horrible, but we knew it was not dangerous, so standard procedure was to drink half a glass before we went out to party. And when we finally left Raufoss, we had a party for everybody we had interacted with in our room. It was successful as we served beer and some real alcohol. Bjorn, who had hoped for some sexual experiences, had bought some pills which when inserted in the vagina, kills the sperm. They were lying on a window sill in a small box, each one packed in silvery paper. The next day I talked to one of the guests and asked how he had enjoyed the party. He answered: "Great party and good drinks, but the candy on the window sill tasted terrible!"

Norwegian technical University (NTH)

Bjorn and I went to Trondheim together in the fall of 1948 to start studying engineering at NTH, the institution that educated all the engineers in Norway. The first problem was to find a place to live which was very difficult. Dormitories basically did not exist in Norway. Even today, in 2015, the newspapers lament the lack of students' places, but of course what they mean is the lack of convenient and cheap places. There was an advisory office run by the students to help place students but it did not work well. But then Bjorn and I had the bright idea to go to the police. They let us sleep in a cell normally used to let drunken people sleep off the alcohol. We contacted the newspaper and they printed the story in a headline the next day: "Students sleeping in jail cells. No rooms available." As a result we got offered three different places for rent, and moved into a place together the next day. It was not a great place, as we had to go through the living room of our host to get into our room, but it was acceptable. We managed to get better accommodations after half a year or so, and moved apart.

I had no contact with Bjorn after that. But we met again 30 years later because he married Inger's older sister Aase. They were both widowed and fell in love. So he became my brother-in-law. Life is not predictable.

The engineering college was arranged such that incoming students went in classes, much like we do in high school in Norway and what is common at universities in the US. This was not the case at the University of Oslo where I much later served as a professor. In the student handbook it is stated that if you, for example study physics, you should contact the physics department in person after about two years! As a student in Norway you have to figure out how to study by yourself with little help from the professors.

As I remember it, about 200 students started at NTH in 1948 and they were all men. The following year another 200 students started, but now there was *one* woman and we got very excited! Engineering has never been popular with women, and it does not matter what women's lib declares, most women just don't want to be engineers. My second youngest daughter Guri went to Brown University to study engineering. That year was unusual, at least at Brown, as it marked the first time that slightly greater that 50% of the incoming engineering students were women. Today this number has dropped and only about 20% of the engineering students in the USA are women. I have had many discussions with my daughters about this. They argue that the differences are more likely to be cultural as opposed to any inherent biological differences between men and women. For example, despite the fact that there are more men in science than women, these gender gaps vary from place to place, demonstrating that cultural factors swamp out biological ones. This may be true, but in both chess and bridge, two games that I know very well, there are tournaments where anybody can participate and men almost always win these, and then there are tournaments for women only. In almost all sports women also do not compete directly with men. For some mysterious reason women and men are declared to be equal, but if you look at the animal kingdom, males and females are seldom equal. It does not mean that one sex is better or more important than the other. But they clearly have different functions in nature as only females can have children. In my opinion this has made women more caring than men, and more patient. Men on the other hand take more risks and are more competitive than women. When Larry Summers, the president of Harvard University, said something similar to this in 2006, i.e. that the under-representation of women in science was due to innate biological differences between men and women, it ultimately led to his resignation.

Very soon I became frustrated with my studies at NTH; I simply was not interested in mechanical engineering. I was, however, an eager guest at the billiard table, and I became quite good at French billiards; I ended up winning the championship at the university. We played for money very often. This was done by putting a few ashtrays on the billiard table. If any ball hit the ashtray, the player had to put in a certain amount of money; in our games this amount was on the order of 1 crown (15 ¢). At the end of the game, the winner had the right to collect the money. I also became an excellent bridge player, and I really enjoyed that game. At one time the electricity was shut off every night at 9:00 P.M. for a whole month in the whole city of Trondheim because of low water levels in the reservoirs. Most of the electricity in Norway is hydroelectric power. Because of lack of light this meant that you basically had to go to bed at 9 o'clock. I borrowed a sophisticated bridge lesson game, made by a famous bridge player named Charles Goren, and played every hand in that game using the light from a candle. My permanent bridge partner and good friend Peder and I won many students tournaments and ended up winning the bridge championship for the whole city of Trondheim. Towards the end of that championship everybody was tired, and Peder dropped some of his cards on the table by accident. The rule then is that the opponents can decide which of the shown cards should be played. This should have been a great advantage for our opponents, but they were also tired, got confused, and asked for the wrong cards to be played.

I also played a lot of chess and won a chess tournament at the university even though as it turned out I did not know all the rules. I learned chess by watching my father play with other adults, and since I was a small boy, 3 or 4 years old, no one bothered explaining the rules to me. I watched the chess game evolve and tried to figure out the rules. This is really very similar to how a scientist works; he or she watches nature and tries to deduce the rules. Nature is not going to tell you what the laws are; you have to figure them out for yourself, or go to college and hope your professors can explain them to you. Interestingly enough, when I was about 40 years old, I bought a computer chess game, and played with it sometimes at night. Then suddenly the computer made what I thought was an illegal move and won the game. I, of course, thought that the program had made an error, and told one of my chess buddies that I would apply for a refund. But he told me that the move was legal, and he was correct. So to succeed in a field you do not have to know all the rules, only the most important ones.

First trip outside Norway

There was a program at the university that allowed students to go to Europe to get working experience when they had been at the NTH for two years. My best friend Magne and I applied to go to France after only one year of study, and we were both accepted. I was very excited about this because at the time it was not common to travel abroad. Magne, however, had been in the Norwegian military and served in the Norwegian sector in occupied Germany after the war. The two other people I knew who had travelled abroad were my parents. My father had a small stipend that allowed them to take a short trip to Stockholm and Copenhagen when they were newly married. Today you have to look hard to find a Norwegian who has not travelled abroad.

I remember I looked out of the train window when we passed the border to Sweden, but of course everything looked just like Norway. It was a tough trip sitting on the train for a couple of days, but we finally made it to Paris. I found Paris very exciting and we visited all the usual places, like the Eiffel Tower and l'Arc de Triumph and of course Montmartre. At night we just walked the streets, looked at the people, and drank wine. We were in France after all! I also found out that almost no one in France spoke English. I had studied French for one hour a week for one year in high school and it did not help me much. Several places in Paris had some games we could play. One involved a bottle of wine hanging on a thread with a piece of chalk in the middle, and by paying a certain amount of money you could try to hit the chalk and shoot the bottle down. If you succeeded you got the bottle. I had an air gun as a child, and was a good shot, so we won that game almost every time we tried. The next day we discovered that it cost more to shoot down the bottle than to buy it in the store.

After a few days of sightseeing in Paris, we had to be on our way. We had two possible places to go. Since this had not been decided in Norway before we left, we decided to draw lots. I was lucky and won, so I got to go to Epernay, the French Champagne district, and Magne ended up in an industrial area in North France, the name of which I can no longer remember. I took the train to Epernay, and upon arriving went to the French Railway office where I was supposed to work. They were helpful and gave me a few addresses where I could rent an apartment. I ended up in the middle of town. The room I rented was on the second floor, and right across the street was the town's big Catholic Church. I rented a room without noticing that the church bell rang every quarter hour, then four rings for every half hour plus the marking of the hours. I did not sleep well on the first night, but I got used to the bells pretty soon. One of the first things the landlady asked me was if I could explain something to her. She had got hold of a small American book written for American soldiers that explained how to pronounce French words phonetically. I tried to explain phonetics to the landlady in my really nonexistent French but she did not buy it because she said: "French is pronounced exactly as written!" The next day I reported to work, and I was even more confused by the French language. I still remember when I said "Bonjour" and they answered: "Ca va", which at the time meant nothing to me. They recognized my difficulties and an English-speaking person offered to give me two lessons a week. It did not work very well because we were both poor in English.

I ate three meals in the cafeteria every day, and we got red wine in a bottle the size of a coke bottle, both at lunch and at night. I liked the food very much, in particular that they tended not to mix the food together; i.e. we had many wonderful vegetable courses, like tomatoes or artichokes. Several young boys roughly my age took an interest in me and we tried to communicate. They got hold of a person who had spent part of the war in a German prison camp and was supposed to speak German. Maybe he did, but it was not the kind of German I knew, and my German was pretty good at that time. When we failed to communicate he said to the other people in French that I could not speak German at all. I understood that and it made me mad, but since my French was not good enough I was unable to object!

I did eventually make many good friends that summer and had a great time. We went swimming often in the evening and swam across the river La Marne to a nice swimming club on the other side. It was for members only, but who knew you once you were in swimming trunks? I also remember July 14 very well, Bastille day. Very often the French arranged dancing in the streets, and on this day everybody came to the main square. This was different from Norway; here the girls came with their family and sat at the same table with their parents. Since

I was reluctant to engage a girl to dance because of my poor French, my friends did it for me. They explained to the parents who I was and told me to take the girls back to their families after at most two dances, which I did. It became a very memorable evening.

At this time Inger worked as an au pair girl in London with a British medical doctor. We wrote each other and decided that she should meet me in Paris and then we should go to Epernay for a short trip. I could go on the train to Paris for free, but I had run into two American soldiers who were driving to Paris that day, and they offered me a ride if I could take them to the l'Arc de Triomphe. I said that would be no problem. I of course was taking a chance, but since "all" the streets in Paris basically radiate from the l'Arc de Triomphe I thought it would not be difficult. It worked like a charm and the soldiers were impressed by my knowledge of Paris. But the real miracle was that I met Inger in Paris. I do not remember where we had decided to meet. It may have been Gare du Nord and we both got there at the same time. If one of us had missed the spot or the time, we had no backup plans and of course this was long before there were cellphones. I remember we were a little anxious to rent a hotel room, because we did not know the sexual mores in France. We wondered if they would ask whether we were married or not, but since we settled on a very rundown place, we figured we were safe and also it was not expensive.

The next day was excruciatingly hot and after we had been to the top of the Eiffel Tower Inger wanted to go swimming. We found out that you could swim in the river Seine. The Seine of course was polluted, but they had a swimming pool-like structure, which was filled with people, and a rather small pool where it was okay to swim. It smelled strongly of chlorine but we had a good time. We spent a day sightseeing in Paris before we took the train to Epernay. The first thing I showed Inger was the sculpture in the city square of a monk, Pierre Perpignan, who is reported to have invented champagne. My landlady did not allow women in my room so we had to rent a room in a hotel for Inger close by. Why

I did not stay at that hotel I do not remember, but it was probably to save money. It so happened that a friend of a friend of mine was getting married the next day, and there were lots of celebrations that night. Basically all my friends made their own champagne and we went from house to house tasting champagne from beer glasses, and Inger was not used to this. Next day I had to go to work, but we agreed to meet at lunch on a small railroad bridge. I waited all my lunch period but Inger never showed up. Waiting there I was surprised that a train that went under the bridge was very quiet and I later learned that this train had rubber-covered wheels. At night when I visited the hotel Inger was barely up and told me she had a very bad disease. It was a disease I knew all too well; she had drunk much too much champagne the previous evening. She stayed an extra day, but then she had to leave to go back to England. I cannot remember whether I followed her to Paris or not.

One of my last nights in Epernay I decided to treat my friends out to a simple meal. While we were sitting at the table a fellow came in and asked for directions to another place. I thought he spoke French with a big accent and said to my friends, "He speaks French badly. Where is he from?" And one of them said, "He is English and speaks French much better than you!" I suppose everything is relative and I had only heard French spoken by native people during my stay. Before I went back to Norway, I decided to make a quick trip to London to visit Inger for a few days. I also served as a guide for my French friends who came along, as none of them spoke English. Since they all worked for the French Railroad they rode the trains for free. We took the subway around London and I remember that my friends enjoyed the trip, but not the food and the English weather. In Paris everybody walks on the shady side of the streets, while in London people walk on the sunny sides. There is much more sun in Paris than in London.

When I was in London I also had a chance to meet the family Inger worked for. They were nice people and told me they had just seen the

Orson Welles movie "The Third Man" and were taken with it. Inger got the night off and we went to see it that night. We parted the next day and I started on my way home to Norway. Of course my first stop was in the industrial city in Northern France where Magne was and we decided to hitchhike to Norway. At the time Magne was in pain because he had arm wrestled with two of his friends at the same time and hurt his shoulder; Magne prided himself on his strength. While Epernay is a beautiful city, where Magne stayed there was heavy industry and it was quite dirty. I had been the lucky one in winning our coin toss. Before we started hitchhiking we went shopping, and I bought a sexy, blue, corduroy jacket with the money I had made that summer. We were paid once a week in cash at the French Railroad in physically big bills because of inflation after the war. They had counted out the pay beforehand and stuck needles through the paper money to keep the money assembled. I loved that jacket and had it for the rest of my college years.

We had an interesting time as we went through the Netherlands, Belgium, Germany, and Denmark. In the Netherlands we tried to buy some milk in a store, so we tried English, "Do you have milk?" No response. Then we tried in German, "Haben Sie milch?"; again no response. Finally in French: "Avez vous de lait?"; no response. So Magne turned to me and said in his Norwegian dialect, "Dom har itte mjolk" (they don't have milk). Then the face of the farm girl brightened, they indeed had "mjolk"! Dutch is in many ways similar to Norwegian dialects; while it is impossible to understand spoken Dutch, I could read many things in the newspapers. It was not too difficult to get a lift at the time, and we soon got to the tip of Denmark and took a ferry over to Norway. This was my first trip abroad and I met my parents who happened to be in Oslo that day and told them all about it. At the same time I went to an exhibition about nuclear energy and the atomic bomb from the USA. I learned that in the future energy would be so cheap that you would no longer need to meter it!

Back at the Norwegian institute of technology (NTH)

I was not very interested in studying mechanical engineering at the university and we had no homework which did not help. We mostly played cards, went to movies, and sometimes dated girls. Once I went to a movie and got mixed up with a girl who had fought with and then left her boyfriend. She then asked me to follow her home. The boyfriend was visibly mad so I was a little afraid he would rough me up on my way home. I walked on a little path and passed by a cord or two of firewood. When I passed by the firewood, a man hiding behind it suddenly sprang out and swung a big wooden stick at my head. He just missed me and fell forward. Without thinking, I hit his neck and he fell flat on the ground. I first ran away, but then I got nervous and went back to see if the guy had been seriously hurt. It was not the girl's boyfriend that I had feared, but someone I had never seen before. He was bleeding from his nose, and was visibly drunk. He apologized to me and said, "Sorry, I thought you were somebody else!" So we parted as friends, and it was the only time I have hit somebody, except for my brother; we fought a lot growing up.

We had mathematics for 2 years at NTH, and as I mentioned, we had no tests or homework. The students are regarded as mature adults even though of course they are not, and are therefore expected to know how much they should study. I did not go to many lectures. I remember when I went to the first class in chemistry and the professor said, "Anybody who is not particularly interested in chemistry does not have to attend my lectures." It was very quiet in the auditorium for a short while, but his statement sounded like a challenge to me, so I stood up and left. I had not studied very much, and wanted to drop out and go for my compulsory military service. I had postponed that until after my graduation. But my friend Kristen talked me out of it and said, "It will be good examination training for you. If you flunk, you can always drop out." In the math examination we had four problems and six hours to complete the test. One problem looked quite easy to me, it dealt with sine and cosine which I knew very well. I started on that problem, but then the math professor appeared and said that since sine and cosine were written with capital letters, it was of course hyperbolic sine and cosine that he meant. I had never heard about those, and so there were only three problems left that I could try. Math had always been easy for me, but this was a challenge. The result of the second year examination was that I had narrowly passed all subjects, while my best friends all failed in one subject each. A rather good student had instructed me and some of my friends in mechanics the evening before that test; in the examination he flunked while we all passed, though not by much. Life is not fair.

Sometimes we had big parties in the Student Union where I often ended up drunk, sometimes sleeping it off on the billiard table. I could rarely afford to buy liquor, because it is heavily taxed in Norway, but whenever someone offered me some, we drank from the bottle, and I was good at getting more than my share. I have always had some problems with alcohol and both my grandfather and my father were probably alcoholics. Since they were pharmacists they had easy access, and after some time my father forbade my grandfather from working in the pharmacy because he would help himself to alcohol. So both my grandmother and my father watched over my grandfather, but only my mother watched over my father. I never saw any of them really very drunk, but every time my father came home from work he took a big drink of whisky before dinner and fell asleep after dinner. My grandfather was also not allowed to smoke either as my grandmother was very strict. My grandfather grew up in northern Norway and loved to fish. As a young boy I often rowed the boat while he fished, then he also always smoked and I had to promise not to tell grandmother. I remember clearly once we went fishing in Mjøsa, and the fog became very heavy, and we were unable to find our way home. I rowed almost the whole night, while my grandfather fished, sang, told stories from his youth, and smoked. When the fog finally lifted we were right outside the dock we had started from.

The last semester at NTH we had to write a thesis, and my friend Kristen and I worked together for a very new professor in refrigeration technology. During this time I got married to Inger. By that time I had been at the college long enough to qualify for student housing and many of my friends also lived there with their wives or girlfriends. It was not very common to live with your girlfriend at that time, so when Inger moved in with me, we got married very soon. We got married at the city hall in a civil ceremony, and Inger's younger sister, Sissel, was the only

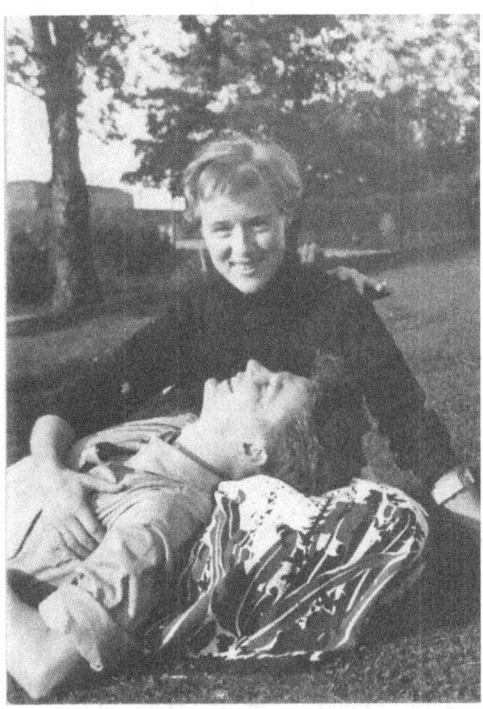

Inger and Ivar engaged in Trondheim, 1952

Married in Trondheim, November 8 1952

relative present. Later that evening we took four or five friends out for dinner to a nice restaurant. I had received some money from my father so we could afford that. We went back to the same restaurant 50 years later and ate boiled cod again, and could not understand what had happened to the intervening years. Living in the student home was okay and we could eat all the meals there. My favorite evening meal was a fried egg, but for that you had to pay extra. Eventually, my friend Kristen and I wrote a thesis together dealing with the efficiency of refrigeration machines, which went very badly for us. The machines had only recently been bought and were in need of repair, so we did not have enough time to run them in order to collect information for our work. We applied for an extension of our thesis, and felt sure we would get it. Unfortunately, we did not and my friend started to write his thesis the day before it should be turned in, while I had started a few days before. My friend got a 4.0, the worst grade you can get and still pass, while I got a 3.5, and so we both barely passed; maybe sometimes life is a little bit fair after all?

ADULTHOOD IN NORWAY

CHAPTER 5

The military

In Norway at the time, one year of compulsory service in the military was required for every male over the age of 18, so rather than seeking a job after graduation, I had to go into military service. Most men do that when they are 18 years old and unattached, but I was 23 years old and married. While I studied I had had a deferment. To me the army was a shock! The first thing I had to do was fill a mattress with hay which was heaped on the floor. Then I had to queue in long lines to get each piece of my uniform, afterwards taking them back to a room in a barracks, with beds for 18 people. The first night a boy froze his foot because he had stuck it out of the window. It is cold in Norway in February. In a speech on the first day a sergeant said, "Whatever you do I don't want to see you standing on the toilets in the latrine." So of course everybody thought there was something wrong with the latrine toilet seats, and we all stood on them rather than sit. The next day the same sergeant asked if anybody could play the piano, and a few people raised their hand. The sergeant continued, "You have to go to the kitchen and peel the potatoes." What I took away from this experience was to never volunteer for anything in the army.

The first three months in the military was basic training, where you learn to shoot, march, shine your shoes, press your pants, and use footpads rather than socks in your shoes. And every evening the sergeant inspected the bed to see that it was properly made, the cabinet to see that it was in proper order, your boots to see if they were adequately shined and if your gun barrel to see if it was clean. I actually ended up paying a guy the equivalent of 5 cents a day for shining my shoes and making my bed. If it was not properly done he had to pay me back twice the amount. One of the first things we learned was that to test whether the room had been properly cleaned the sergeants stuck a finger on top of the door frame, so we always paid special attention to cleaning the door frame first.

While I was suffering in the army, Inger enrolled in a cooking school in a small city, Lillehammer, nearby and rented a small room with a narrow bed. So when I got my first leave I could visit her. It was wonderful even though the bed was narrow — or maybe because the bed was narrow? Anyway the result was the conception of our first child. After three months of really hard training we were allowed to apply for a specialty; I applied to be a typist, thinking I would learn a skill I could use later. After a week the army discovered that I had been educated as a mechanical engineer and assigned me to the kitchen brigade where people with small physical handicaps had been put. I, however, had to do technical work. My first job was to install telephone exchangers in small vans. They had some German ones which were wonderfully designed. Everything was straight and orderly. Then they had some English ones which had crooked, bent metal bars etc., but they worked better than the beautiful German ones.

By now I had moved into a room which had two bunk beds designed to accommodate four people. We were all technical people and members

Ivar in the army with two friends, 1953

of the kitchen brigade. One person was skilled in electrical engineering and rigged up a switch in the door that shut the light off if someone opened the door. The reason was that our light had to be off at 10 P.M. every evening. This was an old barracks and there were plenty of cracks around the door into the hallway. The first evening we were all lying in our beds and could hear the sergeant coming. He ripped the door open, the light went out, and he hesitated for a while, closed the door and walked away. But now he saw the light around the door, came back, ripped the door open, and again it was dark in the room. He repeated this for a few times, and then gave up. After that he never opened the door at night even though the light showed in the corridor. We thought at the time that this sergeant was dumb, but he was at least clever enough to know when he was licked.

Since I was older than most of the soldiers, I ran for and was elected as an "ombudsman" for the several hundred soldiers at the camp, which meant I represented the ordinary soldiers who had conflict with the officers. One such case was a soldier working in a drafting office who tried to be discharged. He did this by filling a wash basin with red ink and visiting several rooms in a barracks at night, waking up the soldiers sleeping there saying, "I want more blood" while dipping a white wash cloth into the red ink. While no one thought he was dangerous I got the army to dismiss him. Another case dealt with a lieutenant that returned to the camp rather drunk and made a lot trouble for the soldiers on guard duty. In the end the lieutenant demanded to be taken home in a wheelbarrow. It turned into a military tribunal of sort, but the lieutenant only got a reprimand as he was the best pistol shooter at our camp and had competed in the Norwegian championship. Most of the complaints I got from the soldiers were really trivial, and mostly dealt with the food. For example, a common complaint was that we were given aluminum cups to drink from, which was okay for cold beverages, but disastrous for hot liquids such as coffee. The cup burnt our hands and our lips, so by the time you could drink hot coffee for example, it was already cold. To be "ombudsman" gave me a certain power, and I could trade favors to get more military leave. By this time, Inger was visibly pregnant and had a job as a receptionist at nearby hotel where I could eat wonderful food for free whenever I visited her.

One day in February the people who went to a military school to become professional soldiers came to the camp. The surroundings were very good for cross-country skiing, and they challenged our kitchen brigade to a cross-country race with full packs. They were all athletes and we were a sorry bunch, but we could not refuse. In the military full pack means that you have to bring all your gear like helmet, boots, rifle, uniforms, etc. and in addition you get a very full pack weighing maybe 50 lbs. The day of the race came and I discovered that the competition had filled their packs with bricks, which means that the weight of the packs is close to your back, and much easier to carry. I found that very unfair, and came up with the following plan. In cross-country races at that time the participants started with three minutes in between, so you were already in the forest, when the next person started. I got most of the kitchen brigade to leave the track, hide in the forest, and just watch the competition hale by. The visiting soldiers had thought that they would easily catch us, and could not understand why they couldn't. Several ended up exhausting themselves by skiing too fast. We were called on the carpet after this stunt, but we said that our ski bindings had come off, so we had to break from the race. Sometimes life is not fair.

On the same day the captain of the visiting soldiers came into the soldiers' mess hall, and asked if anyone played chess. No one answered as we had learned all too well that it does not pay to volunteer in the army. But after a few more challenges, I came forward and said I was just learning and would like to play him. I asked him to explain all the moves to me before we started. As it turned out, the captain was only an average chess player and easy to beat. We ended up playing two games, and he lost badly in the second game. He spoke to the officers in the officers' mess hall the next morning and said he had met a chess genius the day before! The lieutenant I interacted with knew it was me, and we shared the joke.

I spent a lot of time in the engineering office where one of my main "jobs" was to play and beat the lieutenant in a game called "Five in a Row", putting crosses and circles alternately on a paper with squares on it. On the third of December in the middle of a game, the captain telephoned and called me into his office and congratulated me for becoming a father. He said my wife had a boy yesterday and she was in a hospital nearby. He gave me leave for the rest of the day to visit Inger and my new son. It was a very strange feeling. I was now responsible for a little baby. At the hospital Inger was happy and beautiful, but the baby I barely dared to touch. When I left the hospital they said that I had such a rare blood type, so could I please give some blood? I said of course as I was in a happy state. I never reflected over how could they have known my blood type, and later I learned that I have one of the most common blood types in Norway, type O, so I had been fooled.

I remember that I said to myself when I was in the military, "I will never forget how miserable I feel in this place." Now I cannot remember the miserable parts, only the fun parts. Memory is very selective and untrustworthy. Because I served as "ombudsman" I managed to get out of the military a month early, and then I had to look for a real job.

First job

Inger, our new son, and I moved in with my parents, who were also having a difficult time. The reason was that my father had to change jobs. My father had managed a drugstore for my grandfather for nearly 25 years, but when my grandfather died, he could not continue. At that time in Norway the government decided who should run the drugstores, and whenever a drugstore became available several pharmacists would apply. The person who had most experience always won and my father lost to a woman who had graduated half a year before him. He ended up taking over a worn-down drugstore at Eidsvoll which was about 50 miles from Lena where we used to live. This was a great disappointment to my mother as she did not want to move.

Despite the fact that I had lousy university grades I had surprisingly little trouble getting a job. But since I was married with a baby, we still were in need of an apartment. This was impossible to get at the time. It was 1953 and almost ten years after the war, but the socialist government controlled the apartment prices and set them so low that no one wanted to sell. After trying unsuccessfully for some time, I took a job with the Norwegian Patent Office in Oslo because two of my poker friends from NTH already worked there. Eidsvoll was close enough to Oslo, so I could take a bus to work every morning and be back home at night. It took about two hours each way. This worked for some time, but in the end it proved strenuous. I was fortunate to be able to move in with Arnold in Oslo, one of my friends from NTH. We thought that Arnold and his wife Tutta had been very fortunate because they had managed to get an apartment in Oslo. But in order to get the apartment she had to wash the stairs and he had to work as a janitor, so perhaps he was not so lucky after all. During this time Inger moved back to live with her parents in Skreia.

The job in the patent office was easy enough. I was assigned 3 classes: locks, hinges and knitting machines. My mother's wish that I should have been a locksmith was almost fulfilled! I had to examine any patent application in these three areas. A typical application dealing with hinges was this: if a post holding the garden gate had become crooked, then the gate would close itself. So lots of people got the "bright idea" that if you could offset the hinges the gate would close itself. Most people use a patent lawyer when they apply for patents, and when you tell them that the idea is old, the lawyer always tries to tell their clients that their particular design is really different. That is how patent lawyers make some of their money. The knitting machines were mostly simple designs that knitted sweaters if you moved a shuttle back and forth with your hand. In Norway hand-knitted sweaters are more expensive than machine-knitted, and since these machines used muscle power, the sweaters qualified as hand-knitted. I remember one claim from Japan, which I inherited from my predecessor, that was more than 100 pages long and which I did not understand. I managed to object to some small detail a couple of times and sent it back. I suspect my predecessor had done the same thing. So it still may be in the patent office as far as I know. I can only remember one patent which was really interesting from the lock class. Keys to a specific deposit box are flat and can often be filed down from other keys fitting other boxes. The patent was that you only use keys that cannot be obtained from other keys by filing.

Tutta and Arnold who took me in, 1954

I lived with Arnold and Tutta for maybe 3–4 months and I helped them with their chores. In the winter months homeless people bothered us, which came as a surprise to me as I never knew they existed. They usually sat or lay down in the basement close to the furnace or on top of the stairs to the attic. Every evening we walked around to see that everything was okay, and ran into them. Arnold was a very kind man, and usually did not throw them out, but instead told them they could stay on the condition that they did not come back tomorrow. The next day a few different people would show up. So it was a more or less a hopeless problem.

Every day I answered all the advertisements I could find in the newspapers about an apartment for sale or rent, but there were never more than 4 or 5. I always made it a point to say, "Money is no problem." It was of course a problem, but I still got no response. Finally one day in the spring we were offered an apartment in Oslo for two months, not long, but we were very happy. We moved in as soon as we could, and Inger and I lived like an ordinary couple for the first time. By that time we had already been married for a year and a half.

During that time, my bridge partner, Peder, from Trondheim stopped by on his way to the USA, having decided to emigrate. He came with two suitcases made out of cardboard! He stayed with us overnight and we spent the evening together. He would first visit two friends, Magne and Kjell, who had already emigrated to the US. He seemed kind of nervous as he had left his wife and a newborn baby behind, but they were to join him later. Two months later I got a letter from Peder. He had sent for his wife and child, bought a car as large as a tank, and of course he had an apartment.

Arnold asked me one day if I knew that I could register for an apartment in Oslo. I said yes, but the waiting list is 8 years long, and Inger and I were not prepared to wait that long. He said it cannot hurt, and ended up driving me to the registration office. While Arnold waited, I went through the motions and registered. Everything seemed okay until the man behind the desk said, "Where is your wife living?" I said "Right now she lives in Skreia with her parents." "If your wife does not live in Oslo, I can't register you!" the bureaucrat said. In that moment I decided to emigrate.

I had to quit my job in the patent office, and the chief engineer warned me against this saying: "If you leave, you will never get your job back." That did not worry me much. A little before this time I had sent a memo around the patent office that included a calculation on how long it would take for engineers like me to advance and become chief engineers. It was easy to calculate because the dates when the present chief engineers would retire were known and advancement went strictly by the time you had worked there. I pointed out if someone died prematurely, everybody would be happy as they would move one step closer to promotion. It became a much quoted memo. We young engineers discussed it, and I remember one person said that he liked potatoes and sauce just as well as meatballs, so he was happy to wait. Obviously, my character was different.

CHAPTER 6

EMIGRATION TO CANADA

The journey

I emigrated to Canada rather than the USA where all my friends had gone as Inger had a sister living in Toronto at the time. She also had a cousin who had been in Canada during the war, and who was married to a Canadian woman. We visited them before we left Norway and she was nice enough to offer us to stay with her mother when we arrived, at least until I got a job. I felt worried because at the time married women with small kids were not expected to work, and it would be my responsibility to take care of both Inger and my son, John.

We decided to travel by boat to New York, then by train to Toronto. As the Norwegian America Line did not charge for children that were less than one year old, we were in a hurry to leave before John's first birthday, December 2, and it was already November. We ordered tickets on a boat named *Stavangerfjord* which was already an old boat. To illustrate how we badly we were strapped for money in Norway, I remember one time in particular when we were invited out for dinner. Since we could not afford a babysitter we tied our son John to the bed and went out for dinner. I had to convince Inger to go along with my dumb idea, which she reluctantly did. Of course, we were worried the entire two hours or so we were away, and were very happy to find him safe and asleep when we got home. We never did this again, it is clearly illegal and we had risked our son's life for a few dollars. When you are young you sometimes do stupid things. In retrospect I am embarrassed to admit we ever did this, but times were tough.

The boat went from Oslo to Bergen, on to Halifax, finally ending in New York. When we left Oslo my mother and father came down to the docks and saw us off. Strangely enough I cannot remember if I felt guilty for leaving my parents or whether I was even worried about leaving Norway. An obvious motivation for having no reservations about leaving however was the fact that my job at the Norwegian Patent Office paid 12,000 Norwegian crowns a year. It was not difficult for me to figure out that there was no way we could live on that salary, even though I had stopped smoking to save money. This was a hardship for me as I had smoked since I was in 7th grade and continued all through college. Remember this was before we knew how dangerous smoking was, except for my mother of course, who was the first to know! To leave Norway I had to borrow money from my father to pay for the boat tickets, including an additional $200 which was all the money we were allowed to take with us according to Norwegian law.

When the boat stopped in Bergen we were allowed ashore, and Inger and I went around in the streets and looked at Christmas decorations with John in a baby carriage. That evening we finally left Norway and would not return for 5 years. Right after we left a big storm started. We had gotten seasick tablets from my father, and Inger took them but I did not. Strangely I did not become seasick at all on the journey, even though I had been sick every time we drove to some place in a car as a child. Our cabin was at the bottom of the boat, and we had one porthole window. The cabin maid had to fortify the window because of the storm, and she warned us that it should be checked. As I remember the storm was very violent, so the boat had to go around Iceland and it took two extra days to reach New York. On the third day at sea the storm quieted down a little and the captain called us out on deck to show us where the lifeboats were stored!

One morning at breakfast I went by myself to get something to eat, and the breakfast cabin was practically empty. A fat waiter with a very red face was standing by the table where the various food choices were displayed. There was a two-inch high edge built around the table to prevent things from falling onto the floor. Right when I got into the room a giant wave must have hit the boat, because most of the food went over the edge and onto the floor. I fell as well, but when I looked at the waiter he was rocking back and forth with his arms crossed on his chest, as if nothing had taken place.

We never saw many people on the boat until we arrived in New York probably because many people were seasick. There was entertainment on board which we participated in. We were exposed to Bingo for the first time, and Inger won a nice Norwegian skirt even though we only played a few times. As an experienced gambler I knew that the chances are always against you, and we never participated in the card games like twenty one or roulette.

John had not learned to walk yet and was happy to crawl along the floors, but we were not so happy about that because they were not very clean. I spent much of the time trying to learn some English. Normally in Norway you studied English for many years in schools, but the Germans changed that during the five-year occupation and we had German for five years and English for only two years. Also the books we used at the university were in German mainly because the English system of units was and is ridiculously complicated. I borrowed some books at the ship's library and remembered I had read Steinbeck's *The Grapes of Wrath*. I knew the story well, because I had already read it in Norwegian, but there were many English words I had to guess. English grammar is relatively simple, but English spelling and pronunciation I found difficult.

After about 10 days we arrived in New York City. My mother had a friend from her childhood, Rolf Mellerud, who lived in Schenectady, NY, and she had asked him to meet us. He and his wife came to the docks and helped us as best as they could. We had no idea that many years later we would move to Schenectady to live. They became good friends with us over the years and were like grandparents to our children. I remember that life on the docks and in New York City was just as I had expected, busy and crowded with people. Our first problem after disembarking was that since we were emigrating we had several wooden crates with us, and one was missing. We had to spend a long time on the docks to straighten that out, but were not successful. Then the Bennett Traveling Bureau had promised to meet us and hand us the train ticket to Toronto which we had paid for; but of course they were not there. Rolf was a great help and he managed to get a guy with a dilapidated station wagon to take our stuff to Grand Central Station. When the taxi drove off I thought that was the last I would see of our stuff. But I was pleasantly surprised to find out that except for the one missing crate, everything was there. I had to leave John and Inger at Grand Central Station while I traveled around New York City to try to locate our missing stuff. First I managed to find the Bennett Travelling Bureau and collect our railroad tickets to Toronto. To accomplish this I had to travel on the subway, and I had to ask many people for help and directions. To my surprise a couple of people asked me where in England I came from; I had learned English pronunciation in school. One person on the subway told me I had to go to Radio City Music Hall to see the Rockettes, they were fantastic. Memory is strange; how can you remember clearly so many insignificant things? I also managed to find the Norwegian American

Line's main offices and lay claim to the missing wooden crate. I learned that it had been unloaded in Halifax and we were finally able to retrieve it in Canada many months later.

When I came back to Grand Central Station, Inger was exhausted because she had been carrying John the whole day. She had bought a cup of soup for 25 cents and so now we had $199.75 left. She had been accompanied by a woman who claimed she was a clairvoyant and wanted to know our birthdays so she could tell us what would happen to us and to John, in particular, in the future. Since Inger had been an au pair girl in England she had a lot of experience in communicating with people. She told the woman when our birthdays were and received our futures from the woman. Strangely enough the woman said nothing about my future Nobel Prize! Maybe she could not tell the future after all.

To my great surprise, almost as soon as the train had left Grand Central Station there were no lights to be seen. Like almost everybody in Norway, I had expected the entire USA to look like New York City. I discovered to my surprise that the USA is mostly a rural country. Looking out of the windows we saw almost no lights and everything was desolate. I had decided not to eat anything until we got on the train, where we bought two cups of soup for 25 cents each. When I asked what it cost the vendor said "One quarter" which I did not understand so I asked again and the vendor said "Two dimes and a nickel" which left me even more puzzled! To this day soup is my favorite meal even though it now costs more than 25 cents.

Canada

I cannot remember if anybody met us at the station but somebody must have. The next thing I remember was living with Mrs. Griffiths. She was the mother of the woman married to Haakon, Inger's cousin. She was a very friendly woman and welcomed us, two complete strangers, as old friends. But she was going through a very difficult time because her father, who lived with her, was very sick, and it became very clear to us that we had to move very soon. We lived on Dunn Avenue, a street off of Queen Street West. I had a friend who lived on Queen Street, so I decided to find him and ask for advice. I took the streetcar thinking it would take fifteen minutes to get there. About 2 hours later I arrived at his address, he lived on Queen Street East, and of course he was not home so there was nothing else to do but to head back to Mrs. Griffiths. I learned later that Queen Street is 15 miles long and one of the longest streetcars runs in Canada.

We next lived with Inger's sister Aase, but they did not have much room so we rented a room across the road from Aase. Gordon, Mrs Griffiths's son, helped us move. Aase was married to Morten who was an engineering student at the University of Toronto. Inger could then watch Aase's child while she worked, and do some housekeeping. Morten had a small car and I have a vivid memory of one night when we drove through a rich street in Toronto looking at Christmas decorations. Many big houses had Santa Claus and several reindeers on top of their roofs. At the time such outdoor Christmas decorations did not exist in Norway. Many years later Arthur, a good friend of ours, brought Christmas tree lights to Norway and displayed them outside his house. They disappeared the first night.

The worst period of my life

The worst period of my life began when I was looking for a job in Toronto. I was told there was not much point answering ads in the newspapers, so I walked the streets day in and day out, dropping in on places that looked like they would need an engineer. I am a very shy person and do not like to talk with strangers so it was very difficult for me. I had been thoroughly brainwashed in Norway and firmly believed that I had graduated from the best engineering college in the world, and so I thought people would be eager to hire me. When I asked for a job the most common answer was people wishing me Merry Christmas and saying come back in the spring. At the time (and maybe it is still true now for all I know), the Canadian economy was very seasonal, and we had arrived at the very worst time in the midst of winter. I regretted very much having brought Inger and John along and became very nervous as our money was running out. When I said I was from Norway, many people thought it was maybe the capital of Sweden, and what I had thought was the best engineering college in the world, no one, and I mean absolutely no one, had heard about it. It was a rude awakening for me. It was very tough to walk the streets and ask for a job. The pain in my feet was the least of my worries. Before I left Norway I had bought a pair of new shoes. This was the first time I had shoes made from artificial leather. The leather did not stretch, and the shoes were a little too small for me. So I took a knife to them and slit them in the front, hoping no one would notice. One day I had a chance to get a job with Canadian Packers and they gave me a decent interview. But then they asked what I wanted to do if they hired me, and I had no idea what kind of company it was or what they did. It was therefore not surprising that they did not like my answers and I was not offered a job. Canadian Packers is a large food processing company. After that I tried to learn something about the companies before I asked for a job. I then started to research companies in a library before I knocked on their doors. This was a long time before the internet, but I knew something about both General Electric Company and IBM before I went there. When I walked in at General Electric's headquarters in Toronto I came to an office where the secretary asked me what I wanted. So I told her I am looking for a position. She said something like this to me, "Walk left through that door, go to the end of the corridor, turn right and then left again and you are at the employment office." I did the best I could, but I found myself back out on the street. I was surprised and walked a few hundred yards away from the building. But then I hesitated and decided to walk back, as I thought she could not have been that cruel. I made this decision rather arbitrarily, but of course it was the most important decision I ever made in my life. You never know how what looks like a small decision can be very crucial. This time she took me by the hand to the correct office. When I asked for a job they said they were no longer hiring foreigners but if I wanted, I could fill in an application form. I did that and fortunately I wrote that I had some experience from a patent office. The scenario at the IBM office happened much the same way except there I did not lose my way.

What saved us in the end was a Norwegian guy I ran into who knew Morton. He worked for an architect's office. He was a nice young guy maybe 10 years older than me. He arranged for me to meet his boss and he hired me for $250 a month. This was not a lot of money, but since we came to Canada with less than $200 it seemed like a lot to us. We then decided to move and started to look for an apartment we could afford. Unlike Norway there were plenty for rent. I remember one basement apartment we liked at first, but then I drew the curtain away from the window and found that the window was painted on the wall; so much for that place. We finally rented a small apartment on the third floor in Dunn Avenue close to where we had lived before. We shared a bathroom with 8 women who rented the second floor, but nonetheless for us it was seventh heaven. We were on our own, we had a place to live and I had a job, so things were looking up.

To get to my job I had to take the streetcar along Queen Street West and then switch to the new subway going north-south under Young Street. The main problem now was that I had nothing to do in the architect's office and neither did most of the other people. So you had to sort of hide this fact in the big office space. The reason I was hired was that the architect hoped to win the contract to build a big hospital, but he had not succeeded yet. After the first month I was called into the office of the guy who hired me and paid in cash. I got $200 instead of the $250 I had been promised. I am proud of the fact that I did not complain, but I knew right away that the job search had to start again.

The Norwegian guy, who had recommended me, gave me a small sample problem to test my skills dealing with heating and ventilation. I basically knew how to do this work; you have to know the outside temperature, the heat transfer coefficients for the walls, the heating system used, etc. Since the English units proved difficult, I translated everything into metric units and then back again. When I finished after a few days my Norwegian friend said this is not correct, do it again. I did and got the same result, and my experienced friend said again, it is not correct. Then I recognized my mistake; I had failed to translate degrees Fahrenheit to degrees Celsius. Needless to say, it was a frustrating time and I felt tense and upset. I remember once I got home and yelled at Inger. The reason was that she had given away 40 cents to a person who went from door to door collecting money for polio. My wife's father was crippled by polio as a young man and my wife's family had lived with that for their whole life. I understood why she gave the money away, but I yelled anyway because I felt we did not have money to give away. Shortly after that my wife started working for Simpson-Sears, a department store. We found a woman who kept a private child care facility and for a small amount of money took care of John. I often picked John up at her place and made dinner before Inger came home. My expert recipe was spaghetti with tomato sauce and onions. Our son John cried a lot at this time and one day after I had picked him up, he cried in the store and then continued after we got home. I am ashamed to admit that I lost control and started to shake him, and of course he cried even more. Then I suddenly recognized what I was doing and stopped. I have never lost control again, because I always remember how stupid I was at that time. One day we had reason to celebrate because Inger came home with two silver dollars. At Simpson-Sears they had some sort of celebration where they gave out cupcakes to customers and a few of them had silver dollars at the bottom. The girl handing them out to the customers could of course feel by their weight which ones had the silver dollars.

She set many of those aside and gave them to her friends rather than the customers.

One day Gordon Griffiths, the son of Mrs. Griffiths, where we had lived when we came to Toronto, appeared with a letter from GE that asked me to come in for an interview. I had used his address when I applied for the GE job. They were interested in hiring me as a patent examiner. I can no longer remember the interview in detail, but I remember that I said no to the job as a patent examiner. One reason for that was that I knew that patents depend upon the written language only and not on the drawings. I knew my English was not good enough. I also did not like being a patent examiner in Norway; I instead wanted to try to invent things myself. But today it is a mystery to me how I had the courage then to say no to a job with GE that paid $350 per month. Then the interviewer asked if I wanted to start on the "test course." GE hired new engineers, who spent a year or so testing new GE equipment and getting familiar with the company before getting a permanent job. This was an ideal job for me and I accepted on the spot even though the pay was $300 a month, $50 less than a patent examiner. After the interview I went directly to Simpson-Sears and told Inger about our good fortune. I then went to the cafeteria and splurged on a salad for 60 cents to celebrate. It tasted wonderful. Poor Inger was working and could not join my lonely celebration.

Peterborough

The job was in a small city called Peterborough which was roughly one hour or so by car northeast of Toronto, so I could not commute. I quit my architect's job as soon as possible, to start a new and better life in Peterborough. I took the train up to Peterborough and went to the GE employment office. They suggested that I find a temporary place to stay and I ended up staying in a boarding house where 4 or 5 men lived with full pension. It was run by an elderly widow, and I was lucky to be able to get temporary housing. We had no telephone so there were no way I could contact Inger, but I did phone Gordon who gave her a message. I met the other young engineers at the test course, and was delighted to find out that they probably knew less than I did, but then they were a few years younger, too.

My first order of business was to rent a place for Inger and me to stay. I got several leads from GE, but since I had no car it was difficult for me to get around. Then I got a tip from someone about a furnished house for rent for the summer for $75 a month. Why not rent that first and then look around? It was on Amour Road and I could take a bus to GE in the morning. I looked at it and rented it on the spot. It was a delightful, small house with two bedrooms, but one of the bedrooms was closed because the people renting it out used it for storage for their private things. So after three long weeks I could go back to Toronto, this time by bus to fetch my small family.

I did not tell Inger about the house in detail; only that I had rented a place and she would have to wait and see what I had got. I wanted it to be a very pleasant surprise. We had a lot of things to move and were fortunate that Gordon again offered to drive us. I can remember that he had to remove his spare tire from the trunk to make room for everything. He was worried about that but fortunately nothing happened. When we left Toronto we had to say farewell to all of our friends there, and Inger ended up crying, even though we had shared a bathroom with 8 women. When she saw the house I had rented she cried again, this time from surprise and happiness.

Now began a very happy summer for us. I loved going to work to test various switchgears, motors, generators and other electrical equipment. Then I was offered to join GE's famous A, B and C course, which GE had just introduced in Peterborough. It had existed in the USA for many years. The reason for the course was that it was not common for engineers to do a Ph.D. degree at that time and GE wanted to teach their engineers more. Most of my friends did not want to enter the course, I think now that only about 8 out of 24 of us did. This again was a decision that was very important for me and the rest of my life. The course became a revelation for me. We had 4 hours of instruction one afternoon of every week, and were then given a real engineering problem to solve. I found some of the problems very difficult and used lots of time because I chose to work on the problems alone. If you decide to work in a group there is always a smart fellow in the group and you end up copying him. Since I had not exactly applied myself at NTH in Norway, I now recognized that to succeed on this side of the Atlantic Ocean I had to know more. It was very fortunate that I had gotten a second chance to learn, and I did not want to spoil it. We had to write down the time it took to solve the problems, and I often used over 40 hours because I worked alone. I did not want people to know how long I worked on the problems, so I lied in my solutions and said 25 hours maximum. When I went to NTH in Norway I had not been interested in school. One reason for that was that Norway does not really manufacture anything, even to this day. No cars, no airplanes, no TV sets, no clocks, etc. I knew that the probable future for me as a mechanical engineer in Norway would be to run a small sawmill or something like that. To do that you do not need mathematics, you only need common sense, which I believed I had. In Canada, however, things were different. I could see with my own eyes how GE manufactured big generators or small motors, making studying much more exciting. I also noticed at the time that GE management in Peterborough was generally concerned about competition from Europe as some of the equipment from Europe was both better and cheaper. They tried consulting with GE in the USA but got no comfort there. The GE people in the USA made the same mistake that car manufacturers made later: they did not believe that Europe was a real competitor, because they could not manufacture enough volume to challenge and compete in the USA. But since Canada was a smaller market, they felt the competition long before the USA.

We had a very good time living on Amour Road and managed to save some money. Every week Inger went to the bank and put some money into our account. We lived very frugally because I felt obligated to pay my father back for the money he had lent us for the ticket to Canada and the $200 cash we brought. So eventually Inger started to work at Simpson-Sears again, and this time we had a Swedish woman, who was a good friend, to take care of John, and it all went very well. Scandinavians abroad tend to become good friends. I remember the famous boxer Ingemar Johansson whom I used to refer to as a damned Swede, but when he won the World Championship I changed and called him a fellow Scandinavian!

Since I worked together with newly graduated engineers it was easy to make friends. My judgment may not have been the best, but we lent a lot of money to a Canadian couple from Halifax, Dick and Carol, so that they could buy an old car. At the time we had also thought of buying a car, but were not quite ready for it. The deal was that since Carol worked, she should pay us her salary every week. For a while that went okay, but then she did not want to do that anymore. I insisted, and fortunately they finally gave in. We lent them the money because they were really good friends and a very helpful couple and we learned a lot about life in Canada from them. Also Inger and I tend to trust people and fortunately we have never really been disappointed.

The work at the GE plant was very interesting to me. We tested very large DC motors and here you had to be careful because they could pick up very high speed if connected to the wrong voltage. I also measured the efficiency of AC motors and the operation of switchgears. Because it was necessary to use high voltage you had to be careful. Very often you were told to work with your left hand in your pockets to minimize the chance of getting a current though your heart. Once I shorted 600 volt with my screwdriver and it melted and that caused me to sit down for a while. Sometimes I found logical errors in the switchgears and that became one of my specialties.

After that summer we had to move, and we found another furnished apartment on the third floor in an old house. I remember looking to see how we could escape in case of fire, and instructed Inger about what we should do. We did not like to live on the 3rd floor with a small child. Life was not so pleasant anymore because the apartment was really kind of dirty, the furniture was rickety, and the landlady kind of grumpy. However we celebrated Christmas with a small tree and gift for John. Sometimes that winter I had to work overtime on the night shift. I liked to do that because I got extra pay. But on the other hand I did not like it because in the middle of the night I had to play cards with the other people on the shift. This unofficial rule was not negotiable and because this was unacceptable by the management I was afraid I would lose my job if GE management found out. But since it was a custom on the night shift I had to do it to keep on the good side of my friends. It reminded me of Norway where I could not make as many pieces in the foundry as I wanted. I guess every country is similar in this way, with some strange customs.

Another scare we had that winter was that John became very sick and could not keep any food in his little body, so he ended up in the hospital. For some strange reason we were not allowed to visit him; they said it would be much better if we did not. Of course it was impossible to keep Inger away from her sick child, so we went to the hospital and peeked in the door where John was. It must have been serious because he was being visited by a priest at the time, but maybe he just tried to convert him? John suddenly saw Inger, both of them started to cry, and we managed to give him a good hug before they threw us out again. I never did understand why the hospital had this policy. I remembered the stay I had in the hospital when my tonsils were taken out but I believe the reason I was left alone was the distance and time required to visit and not the hospital's policy. About three years later when we were in Schenectady, I took John to the GE laboratory where I was working to attend a Christmas party. The corridors are green, long, and can remind you of a hospital. John and I were dressed for winter, and I said to John, "Why don't we take our coats off here?" and he started crying. He obviously thought we were back at a hospital.

In the spring we moved back to the house in Amour Road and I graduated as the best student in the A course, and now I had become interested in learning more. But unfortunately there was no B or C course in Canada; I had to move to Schenectady, NY. And we began preparing to emigrate to the US. I knew the Norwegian emigration quota was not filled, and when GE asked if they could help me in getting a visa, I foolishly said it was not necessary. Well, we applied for a visa but heard nothing from the American embassy in Toronto. So I wrote and asked what the holdup was and soon received a reply. It was a form letter with about 25 answers, 3 of which were crossed off to supposedly answer my questions. The 3 answers had of course nothing to do with my question so I wrote again and the same scenario repeated itself, but now 3 different answers were crossed off. I went to the personnel manager after the third attempt and asked GE for help, and about two weeks later we got the visa. Bureaucracy exists everywhere even in the USA!

I contacted GE in Schenectady, and they said I could not be transferred but if I showed up, they would probably hire me for $400 a month. I had now gotten a $50 raise in Canada and so everything sounded fine with me.

I knew the personnel manager at the factory, as I had recommended that he go to Norway to hire engineers and he did. But he was almost arrested in Norway, because of an advertisement where he said GE was willing to pay travel up front, and they could pay it back later. This is illegal in Norway according to an old law, but fortunately he did not blame me for it. I asked him if there was any way I could be transferred, but he said it was against GE's policy. The reason was the Korean War. If young Canadian men went to the US they would be drafted, and no one wanted that. If you were married and established in Canada, moving to the US was generally desirable because the pay was better. If you were too old to be drafted, quitting your job was generally considered too risky, and to keep engineers in Canada, GE did not transfer people.

We had to move out from the house in Amour Road when the owners came home from their summer business venture. We managed to rent a room for one month in a house very close to the railroad track. A train passed by at midnight, and both Inger and I jumped out of bed; it sounded like it came right through the room. But a couple of nights later we slept right through; you rapidly get used to regular noise. By now I had received a message that the B course, which was very mathematical, was about to start and I could come down to Schenectady for the day and participate. I was delighted and arranged to take the train down, and to stay in a hotel. The B course was an awakening for me. Two people were from Cooper's Union in New York City where the best people can go without paying tuition, and I was amazed by how much these people knew. They were simply unbelievable to me. I had little difficulty competing with the Canadian engineers; this was going to be an order of magnitude more difficult. I was given a set of problems in Schenectady and now I tried unsuccessfully to solve them.

The following week I was to go down to Schenectady again to continue the course and I was very nervous because even though I had worked very hard, I had not managed to solve more than half the problems. Inger felt that I had neglected her and John, and I admit I did. This is the only time I can remember that Inger complained, even though I became incredibly busy the next few years. This time when the train got to the US border, the border guards threw me and an Italian family off the train, suitcases and all. We did not have the correct visas. This was before the train reached Buffalo. The Italian family consisted of a wife and a husband with three kids under the age of 10. We had to walk along the railroad track towards some light we could see in the distance. I helped carry their suitcases, while the husband was swearing and the wife was crying carrying the youngest child. The light turned out to be a Canadian bar and we all rented rooms for the night. I was furious, as I had entered the US the previous week with no problems and did not believe I had any visa problems. That night I did not sleep much as my room was directly above the bar, and Canadians are not exactly quiet drinkers.

The next morning I took the train back to Toronto, and went directly to the American Embassy to complain. I explained my problem and my anger to the guy behind the desk and tried to persuade him that the border patrols had made a mistake. He insisted that I had a transit visa, the officers had made a mistake when they let me enter the US the previous week, and there was nothing anybody could do. I tried to argue some more, but he interrupted me and said: "If you don't stop I will charge you with illegal entry last week." I knew I had lost, went out and bought a rubber rabbit for my son, something else for Inger, and returned to Peterborough. My impulsive rubber rabbit purchase was a big hit with John and it became his favorite toy for many years. He was not spoiled by gifts, as we always saved money, because we had little security. Inger was surprised and also very disappointed when I suddenly appeared at home a day early and when she heard the story. Two weeks later we were told that we could pick up the visas at the American Embassy in Toronto. In the meantime I was offered more money to stay in Canada, but my mind was made up. We packed what we thought we could take with us. Inger wanted to take a small rocking chair we had bought for John, but I thought it was too much. I lost that fight. By this time I had thrown away most of the useless things we had brought from Norway, but even so, we still had a lot of things.

Fortunately for us the personnel manager offered to drive us to the train station. He helped us with the luggage and waited until the train started to move out of the station. Then he called out, "Ivar, you are transferred after all!" I became elated, because I no longer had to worry about whether I would get rehired or not. How I got the courage to quit my good job in Canada I cannot understand now. It must have been because I was young and foolish but also because I enjoyed learning new things.

We had to stop at the American Embassy to pick up the visas; our Norwegian passports were still valid. When we arrived, we stated our purpose for being there and then we had to wait for a very, very long time. The whole embassy was watching the last game of the World Series, and now that I have lived in the USA for more than 50 years I finally understand that baseball is an important game. So that day we had plenty of time to look around in the office. Hanging on the wall was an award proudly declaring that they had been voted as the most efficient US embassy in the world. This was measured by the number of letters they answered compared to the staff. I could attest to that, but there was no category for the effectiveness of the letters! I had personally received three of those meaningless letters where they just crossed off some answers. So in this case at least, efficiency was relative!

CHAPTER 7

EMIGRATION TO THE USA

B and C course

We did finally get on our way. When we reached Schenectady we already had a hotel reservation at the Van Curler hotel, an old hotel that has now become part of a Community College. It was a very nice hotel at that time and we went down for dinner. GE paid for the stay and to celebrate our arrival we ordered lobster. The waiter said something we did not understand, so we repeated our request. We later learned that we had arrived on Thanksgiving and everyone, of course, ate turkey. This was a surprise to us as we had celebrated Thanksgiving in Canada a month earlier. It falls in October in Canada and November in the US. Inger told me later that our second child, Anne, was conceived at that hotel.

One of the first things I had to do was to show up at the A, B and C course office to get an assignment. The normal day-to-day life for a person on the B course was to work alongside a senior person for 4–6 months and then switch to work with a new senior colleague. When I got to the office they showed me a list of people and told me who was easy and who would be difficult to work for. Being brash I picked the most difficult person, Dr. Hillel Poritsky, a famous Hungarian mathematician. I picked him because I reasoned that if he was difficult to work for I would learn more, which I did. I also met one of his assistants, Peter, who worked there and he became a lifelong friend.

On the home front we still had to rent an apartment and buy some furniture. We looked around some, and since we did not have a car, we could not look too much. We settled on an apartment complex called Daisy Lane which had about 50 apartments altogether. Fortunately for us it turned out that this was a temporary stopover for many people starting at GE, and again we picked up many lifelong friends from there as well. I could get a ride down to the GE complex in downtown Schenectady with an engineer, Bud, who lived on the first floor right under us, which was very convenient. And now we were ready to buy a car, and Bud helped me pick up the car. We had decided long ago we wanted to buy a Volkswagen Beetle, because it was the cheapest car you could buy. We paid $1730 splurging on an added sunroof that we enjoyed very much. Bud and I picked up the car in Glen Falls about 50 miles away; he insisted that I drive it home. He sat beside me and gave me advice. Since we now had emigrated to the US I had to apply for a driver's test to get a NY license. I already had a Norwegian license, but had barely driven. I thought it best to take a few driving lessons before I tried for the license. I failed the first time, because I had to side park. Although that went well, when I took my foot off the brake the car on its own inched forward and gently hit the car in front. I had of course used a gearshift on the Volkswagen and had to engage the gear to move the car, so I was not prepared for the automatic motion of the test car.

B-course

At work I was very busy solving mathematical problems, and fortunately I found a German book in the library where many linear and nonlinear equations were solved. So whenever an engineer contacted me for help, I would ask him to come back the next day so I could have some time to look at the problem. Then I looked through the book and could normally find a solution. The B course also took a lot of effort. While most students worked in groups, I again chose to work alone. When you work in groups there are always one or two people who do the actual work, while most people just copy the results. It took roughly 40 hours of work to do the problems, and lots of time had to be spent in the library. I enjoyed the work, but I have never worked so hard — either before or since. In addition to the car we also bought a TV set, mainly for John and Inger. I worked every night and allowed myself to watch one half hour of TV every week. The show I watched was "Have Gun—Will Travel".

When I finished my work with Dr. Poritsky, he took me, Peter, and a couple of other people working for him out to a farewell luncheon in a famous German restaurant, Nicolaus, in downtown Schenectady. It was a very nice luncheon, but I remember it mainly because we all split the bill and I had to pay for my own lunch! But I enjoyed working for Dr. Poritsky and learned a lot of mathematics, in particular what kind of problems GE engineers had to face.

After that I worked for Phillip Alger. He was known as Mr. Induction Motor in GE and he was a very interesting person. My English was still not very good, and Mr. Alger had a strange speech defect: he never finished his sentences, so I had a lot of difficulty understanding what he wanted from me. While working for him I developed an equivalent circuit model for an induction motor, which became my first technical publication. He sent me to an electrical engineering (IEEE) meeting in New York City to give a technical talk. I was petrified before giving the talk, and still remember the guy who talked before me. He was relaxed, made a few jokes and used no notes. I, of course, delivered the talk by mainly reading which is a cardinal sin when you speak to an audience. Mr. Alger also wanted me to join the IEEE which I did.

Alger had an early computer in an office I shared with Lisa, a female programmer. I do remember that while I was working for him, the computer was upgraded from a memory of 1024 to 2048 bytes! I helped Lisa program the optimum size of induction motors depending on its rating. One result we got for a 50 horse power motor was that the rotor should be 2 inches in diameter and roughly a mile long! We had to change that program! I became good friends with Lisa and told her that I had played quite a lot of bridge when I went to college. She introduced me to Joe who was looking for a partner. He was the best bridge player I have ever met, and I could not understand why he was looking for a partner. We played in about 6 local tournaments that winter, winning five and finishing second in one. In one tournament we won a wonderful trophy inscribed by Charles Goren, a bridge payer I had admired while I was still in Norway. The trophy was a naked man running with a deck of cards in his hand so when I proudly came home with it, Inger unceremoniously put it under the bed.

We also played once a week for money with a rich couple from Amsterdam who were friends of Joe. They were addicted to the Groucho Marx TV show "You Bet Your Life" so we always stopped for half an hour to watch it. We were then served drinks by a black maid as these people were very wealthy. An example of that was that they bought a Ford Edsel sports car and kept it for two weeks before they sold it back to the dealer. They found it too difficult to both enter and exit the car. They loved to play bridge but their skills were only about average. We made about $50 each night depending on the cards. When they lost the host always took a roll of money with new bills from his pocket and paid us. When we lost, which was not often, we paid with ordinary bills, which we put on the table. I never saw them touch the money. I suspect that they just left them on the table as a tip for the black maid.

After playing with Joe for one year I did not sign on again, claiming I was too busy with school. In truth, he was incredibly rude when we played in tournaments and I was embarrassed to play with him. Most people play in tournaments for fun, and many do not really know the rules in a bridge tournament. For example, you are not allowed to hesitate when someone tries to finesse say a queen. If you have the queen it is okay, but if you don't have it, it is a no-no because if you hesitate then your opponents may think you have the queen. In almost every game Joe shouted "Tournament Leader" and explained some infraction against us by the opponents. Most of our opponents really had no understanding of what they had done wrong, and therefore I found his behavior upsetting and rude.

Mr. Alger also had an interesting secretary I talked to quite a bit. Every time Mr. Alger was travelling, which was quite often, she worked on the file cabinets and threw out lots of files. When I asked why, she said she had to do it when he was out of town, because he would not permit it. She also firmly believed that she and Mr. Alger had been married in an earlier life, and there was nothing I could do to convince her that it was not possible. With her sister she supported an average landscape painter, and gave me a picture to take to Norway. This was not a good investment for them, because several years later I got a hand painted Christmas card from the sisters! The painter had clearly not been a financial success!

Anne is born

About this time we were going to have our second child, Anne, and I decided to take my vacation which was one week, to help out. John was now 5 years old and needed somebody at home. I drove Inger to the hospital when the time came and left her there. They told me they would call when. something happened. At the time once the father had dropped the mother off at the hospital, he had completed his job and was never present at the birth. The birth went well and a few days later I could pick up Inger and a baby girl at the hospital. Unfortunately, a few days later Inger became sick, and I sent for the doctor who actually came to our apartment. He calmed me down, said Inger had a slight cold, and that the baby would not be sick. He told me that all babies had a functional immune system at birth because they get antibodies from their mothers. When he left he said, "You are the only person I feel sorry for in this household!" As a result we kept on to this doctor for the rest of his practice. After arranging for diaper service etc., I went back to work with Mr. Alger.

C course

In my next assignment I worked for a Hungarian applied mathematician named Gabriel Horvay. He was a very interesting man and my first job was to check the accuracy of his calculations. Again I shared an office with a woman and we worked in parallel. We looked up sine and cosine tables with six decimals and used Monroe calculators to evaluate the formulas. Every so often we compared results and if they did not agree we figured out where we had made the errors. At the time Dip Forming had just been invented. The principle behind this is that when you pull a thin rod of, for example, copper through a bath of molten copper, you can get a thicker or thinner rod depending on how fast you pull it through. Dr. Horvay suggested this problem to me, and I had fun working on the mathematics. I used a digital computer to evaluate the results. It was not difficult to program because it worked the same way as an analog computer. Engineers at GE were used to analog computers and someone had made a digital computer simulate the analog computers they were used to. Dr. Horvay asked me to write my findings into an article and so I wrote a short paper. But he did not 'appreciate' my writing and rewrote the paper in a very long form. We sent the paper in to a journal, and it was accepted, but the paper was too long, and had to be shortened. I of course agreed with that and felt a little triumphant because my version had been short. But Dr. Horvay simply retyped his version with single line spacing instead of the double spacing in the original version, and this time it was accepted!

Dr. Gabe Horvay worked at GE Research Laboratory and working for him was my first meeting with this organization. To me it looked like a wonderful place to work. I had no prior idea that it was possible to get paid for just doing research. I count this fact as one of my biggest discoveries! Watching the people there, I saw that they were discussing problems together, chatting, sitting by window sills or writing equations on blackboards. I decided if at all possible I would try to get permanent work there, but I knew that everyone there had to have a Ph.D. degree.

My final assignment on the C course was with a famous German mathematician, Dr. Hans Buechner. He worked for the Steam Turbine department, which at that time at least, was the biggest building in the world. They had a very big crane which reached from wall to wall and ran the entire length of the building. I had a small office at one end, and could hear the crane when it approached. At the beginning I was afraid it would not be able to stop and would crash through the wall, but it never did. The problem Dr. Buechner wanted me to work on was to figure out stress in a notch in a rotor. He had a special way of doing this, using complex variables. First we spent a week trying to estimate the answer. Then we started the real calculations on paper. We used large notebooks, twice the size of ordinary ones, and calculated, which took about 6 months. He was very thorough, and we wrote the equations down on paper with many parentheses preceded by square roots of 2, plus or minus signs etc. Every Friday afternoon we compared results. If the results did not agree, he calmly took my papers and said: "Now let me see where you have gone wrong." I was always wrong, and he was always correct! It drove me nuts, but however careful I was, when we did not agree, he was always right. Dr. Buechner was a large man and he chewed gum. He also impressed me doing that, because when he opened a pack with six pieces of gum, he always ate all of them together. When we finally finished the calculation we had a neat formula to evaluate, and Dr. Buechner asked me to evaluate the result using a digital computer. This was my first attempt at programming. Part of the personnel in the building was about 30 women who were responsible for numerical calculations. These women were computer programmers, i.e. they knew how to program computers in machine language. Three of the women had just gotten back from a short course at IBM where they had learned FORTRAN. In the Steam Turbine department there were two IBM 650 computers they used. When I asked them to help me and teach me FORTRAN, they were very willing to do so. But they warned me that I was wasting my time. FORTRAN, they said, took too much computer memory and would never be practical! Nevertheless they were willing and lent me a small booklet they had gotten at IBM. At the time we used punch cards to write programs and I remember my first attempt at programming. After first sending my cards through the compiler, I got lots of paper back. I started reading the text and it told me I had made an error. Reading a little more it said that I had made a second error. When I got a little further in it said, "You have made three errors. Are you an idiot?" and on the next error it said: "You are an idiot!" I felt very ashamed because I thought everybody must know that I had made at least four errors. But with more help from the women, the program eventually ran successfully. As soon as I put the program in the computer, the big tape wheel made a small angle turn, and I started to walk away. But the guy who ran the computer said: "Don't you want your results?" It had taken only a couple of seconds to get the answer. I understood then what a terrific invention the computer really is!

Working for Dr. Buechner, I was impressed by his great skills, and depressed as well, since I felt I could not compete with him. And since our results from the computer calculation were only about 10% different from the early estimate, I told him I wanted to go into physics next because we do not understand the strength of the material to the 10%. He warned me about changing to physics and said mathematics would be easier for me. I said, partly as a joke: "But you are not in physics, so I do not have to compete with you!" He took that comment with a friendly smile.

While on the C course I had the opportunity to take lessons in giving speeches, and it was an excellent and very useful course. Scientists and engineers in general do not recognize how important it is for your career to be able to clearly explain your work. This is in part because no one tells them and no one teaches them how to give a talk. Of course if you are very good scientist like Professor Hans Bethe, you might get along well without that skill. I once heard Professor Bethe give a talk at Cornell, and it was very boring. The audience did their very best not to fall asleep. But then Professor Bethe said: "This reminds me of a good story" and the audience sat up expectantly. But then he continued: "But I don't have time to tell it now." The audience sank back into their stupor. In the speech class we were supposed to prepare and give three minute talks every time we met. Then we were critiqued by the teacher and classmates. In the first class the instructor had to stop a speaker after twenty minutes; he had no idea how much time three minutes were. I had just gotten my Volkswagen and gave my first speech praising the Beetle, or so I thought. I remember that one person in the audience said: "Thank you; I had plans to look into a Beetle for my daughter. You convinced me not to buy it!" So I failed miserably in my first speech as well.

At that point, I had already started to take some courses at a nearby college, Rensselaer Polytechnic Institute, and enrolled in a master's program in physics. I went back to Dr. Gabriel Horvay and asked if he had any suggestions for how I should go about getting a job at the GE Research Laboratory. He said ask Dr. John Fischer, who was a branch manager and was known to take risks. I got an appointment with Dr. Fischer and learned that he was no longer a manager; he had gone back to do research. But he said Dr. Roland Schmidt had taken over his management job, and he would ask on my behalf. Dr. Schmidt arranged for me to interview with several scientists, as is normal when seeking a job. I also had to go to the personnel department, and they wanted to see my examinations papers form Norway. I had very bad grades from Norway and dreaded the thought. But I had no choice but to dig them up. My luck was that they were all in Norwegian and I had 4.0 in both Physics and Mathematics. In Norway 4.0 is the worst grade you can get and still pass the course. In the USA of course 4.0 is the very best grade you can get. When the personnel director saw my grades he said: "I see you have 4.0 in both physics and mathematics, you must have been a very good student!" I pride myself on being an honest person, but I did not think this was the right time to explain to the personnel manager the difference between the grade system in Norway and USA. But of course they also asked the opinion of the four people I had worked for in Schenectady. I learned later that Dr. Horvay gave me the best recommendation I have ever gotten. He simply said: "Giaever is better than Horvay!" As far as I know I am the only person who has been hired on the staff of the GE Research Laboratory without a Ph.D. degree. I have to add that I was originally hired for a one year trial, but later transitioned into a full-time job.

CHAPTER 8

NORWEGIAN VACATION

To get a job at the GE laboratory without a Ph.D. was a good break for me, so Inger and I decided to go back to Norway for a visit before I started. We had now been away from Norway for about 5 years, and we could afford to make the trip. We flew with Icelandic Airlines because it had the cheapest airfare and a boat would be too slow. I told the laboratory that I would take two months leave, and start full-time when I came back, and that was okay. I also asked what I would be working on and they said "probably thin films". So I took a book on photography with me to Norway, as that was the only film I knew, but as I would later learn, they had very different films in mind!

We were fortunate to get a friendly couple to drive us to Idlewild Airport which was renamed Kennedy Airport right after Kennedy was shot. We managed to store our furniture for two months in Daisy Lane, so we did not have to pay rent when we were away. I still remember how impressed I was when we drove over the Tappan Zee bridge, and then again when we arrived at the airport. Our son John was about 6 years old

Inger and Ivar, first trip back to Norway, 1958

then and our daughter Anne was 2. We got a good deal with Icelandic Airlines: they offered us a free one-day stay in Iceland, or so we thought. When we got to the waiting lounge at Idlewild, we saw only passengers and no officials. Then some guy showed up, put a uniform-like cap on his head and announced that the plane would be an hour late. He then removed the cap so he could no longer be recognized among the passengers. This was repeated at least four times, but we finally took off. At the time we flew to Gander Airport in Canada to refuel before we landed in Iceland.

It was good for Inger and the kids to have a stopover and we had a chance to see Iceland. John had very blond hair, and we were used to keeping track of him by his hair color, but almost lost him in Iceland because everybody had blond hair. Iceland was settled by Norwegians roughly 1000 years ago. If you ran into trouble with the law in Norway at that time, the penalty was to be declared "fredlos" (no peace) which meant that anybody could kill you with no penalty. For these people it was better to take their chances and settle in Iceland. Iceland is sometimes recognized as the world's oldest existing democracy, as its "Altinget" (congress) was set up in year 930 and still exists today. In Iceland we took a bus tour and saw the prime minister's residence, which could not really be distinguished from all the other houses. I had learned a little Old Norwegian in high school which is somewhat similar to Icelandic. I tried some of the phrases I remembered but could not make myself understood. We learned from the people running the small hotel that they had not seen the sun for three weeks, that the cars drive on the left side of the road, and that there were no potatoes to be had. Iceland is a strange place with only a few hundred thousand inhabitants.

When we went to board the plane again the next day, at first they did not let us on. When I showed them the agreement they said it was meant that I should stay for one week, and pay for only 6 days myself. After some discussion they gave in; again I was lucky and we headed for Oslo. At the time it was allowed to smoke on airplanes and Icelandic Airlines was known as the marijuana airlines. Lots of young people were smoking weed in the back of the plane, and the second-hand smoke could be clearly distinguished from the regular tobacco. My sister, Mette, who is younger than me by 9 years, met us in Oslo and took us to my parents who still lived in Eidsvoll, one hour away by car. Since we had been away for 5 years, we had a lot of catching up to do. And we were of course proud to show off our two kids to their grandparents.

My mother and father enjoying our kids, 1958

We stayed in Norway for two months, rented a car for a week, and left the children with my parents while we drove to see the spectacular fjords on the west coast of Norway. We had never been there before, but, ironically, many people in the US had told us it was worth a trip, and it was. The brief trip to our home country went by very fast and we returned to the US somewhat hesitantly. When we got back John started kindergarten and became very unhappy. Before we left for Norway we had spoken Norwegian at home and English outside. So John spoke Norwegian to us and English with his friends. But when we got back to the USA he was unable to distinguish between the two languages. He understood when someone talked to him in English, but he answered in Norwegian. So of course he was made fun of and became unhappy. To remedy this Inger and I decided to speak English all the time. It is very difficult to change the language you are used to speaking with someone but we managed. When we went to bed after the first trying day Inger said: "I senga snakker jeg norsk!" (In bed I speak Norwegian.)

GE Research Laboratory

CHAPTER 9

When we got back I looked forward to starting my new job. It turned out to be very difficult because no one told me what to do. When I asked my manager he said: "Good work", and left it at that. In those days, Ph.D.s that were hired normally continued with their thesis work in their new job. In fact many scientists continue to work on their thesis throughout the rest of their career. I still did not know what to do and went in desperation to the department head, but he also said "good work." Then I went to Dr. John Fischer and he asked me what I wanted to do. I said: "I want to learn some physics but I know I am too old to have an impact on the field." I knew that most physicists make their discoveries very young and I had now become 30 years old. John said, "That is not true because when you are young you are learning things. Since you do not know any physics, you will acquire new knowledge and can make discoveries. My advice is to change careers every 5 years or so, then you can make discoveries your whole life." John was a very wise man. Then he explained that he was also a mechanical engineer like I was, but he had worked in physical metallurgy for several years. He had resigned as a manager because he had some ideas about fundamental particles. He was going to apply second quantization which was an approach used in the theory of superconductivity. I did not understand what he said, but I was impressed. Then he told me that he wanted me to work with thin films, and he wanted me to place two metal plates very close to one another so that electrons can tunnel from one metal to the other. I had never heard of tunneling and must have looked perplexed so John continued, "If you throw a tennis ball against a wall it will bounce back." That I believed. But then he continued: "However, if you throw it enough times, sooner or later, the laws of physics permit the ball to go through the wall and appear on the other side." That I did not believe. But he continued, "The ball will not have changed and there will be no hole in the wall." Now he left me incredulous. So he said, "I want you to try to place two metals about 20 angstroms apart (about 2 billionths of a meter!), and see if you can measure this effect." I remember I thought that maybe this was a theory in physics, but that it would never happen in real life. But since they were paying me I had to please them and give it a try. And now I had something to do, even though I got no more instructions.

I started to try making thin insulating films, and the first thing was to try Langmuir–Blodgett films. It turned out that Dr. Blodgett still worked at the Laboratory and she was delighted to show me how the films were made. Surface active agents (surfactants) are a larger class of molecules. Generally these molecules consist of a hydrophilic (water soluble) and a hydrophobic (water insoluble) part. This two-sided or 'amphiphilic' nature of surfactants is responsible for their behavior in solution. They spontaneously form membranes like micelles, bilayers, vesicles and your ordinary soap bubble in solution and also accumulate at interfaces (air/water or oil/water) with one end of the molecule in the water and the other end in the air. The hydrophobic part usually consists of hydrocarbon or fluorocarbon chains while the hydrophilic part consists of a polar group. I ordered the workshop to make a tray and went to work. The first thing I did wrong was to make the tray much too big so it became very heavy when filled with water. This caused me to spill water all over my work bench. I tried to deposit the films on small flat copper pieces and I had no problems with that. The thickness of the soap film on copper depends only on the number of times I dipped the copper piece into the liquid. Then I put mercury drops on top of the films and arranged a simple measuring circuit using an ammeter and a voltmeter. I used a battery to supply the voltage and a variable resistor to control the current. I continued with this setup for some time and showed Dr. Fisher my results. I remember he called them miracles because miracles happen only once, and I could never reproduce my experiments. I started to believe that the liquid mercury that formed the counter electrodes would diffuse through the Blodgett film and eventually short out the experiment! I told Dr. Fischer that, but he did not believe me. I had great respect for Dr. Fisher, but nevertheless I started to look around for different films. I soon hit upon some papers that described evaporated aluminum films and the non-conducting aluminum oxide which form on them. That seemed like a reasonable idea. But I did not have an evaporator and had never used one before. As usual I talked to Dr. Fischer and he said simply: "Order one." I worried because it cost over $2000, which was more than I had paid for my car. Dr. Fischer put me somewhat at ease by saying that GE hired me to work for them but if I sat in an empty room, I certainly could not justify my existence; I need something to work with. It took about two months before the evaporator arrived, and I spent the time educating myself about, among other things, aluminum oxide. Then I came across one paper that claimed that the oxide conducted! I had made an error ordering the expensive evaporator. I showed Dr. Fischer the paper and he said: "They are probably wrong". It had never occurred to me that a scientific paper could be wrong, and I hoped John Fischer was correct.

Ivar displaying a tunnel junction, fall 1960

Tunneling in superconductors

When I finally got the evaporator, it changed my life. I had always thought I would be a theorist as by now, I had extensive background in applied mathematics. But I loved the evaporator and found experiments very satisfying. It did not take long before I had completed credible experiments showing that indeed I observed electron tunneling through the aluminum oxide. I made the tunnel junctions by first evaporating a millimeter-wide strip of aluminum onto a microscope's glass slide. Then I exposed the glass slide to air for a few minutes and a thin oxide would form on the aluminum surface. I next evaporated a thin strip of aluminum across the first strip, and now the two aluminum strips would be separated by a thin oxide layer. I had learned enough quantum mechanics at RPI to understand that tunneling was possible. The solid evidence I was able to produce was

Ivar posing at the evaporator, Oct 23 1973

that the resistance was inversely proportional to the area, was almost independent of temperature, and for thicker oxide, was nonlinear with applied voltage. I tried to write a paper about all of this, and showed it to my friend Walter Harrison. He told me that I had stolen too much from a paper by Ragnar Holm, a German physicist who had written a book about electrical contacts.

I was then asked to give a talk at the laboratory about what I had done. I remember being very nervous: here I had to give a talk to a room full of people with Ph.D.s, and I had barely started to study physics at RPI. But fortunately for me one of the first problems students learn in quantum mechanics was tunneling, and I recognized now that it was a real phenomenon. I prepared very well. I did not read the talk, showed some slides and was happy when the audience clapped. But then the talk was opened for questions. I got several questions: "Is the current due to semiconducting?", "How about small metal bridges?", "Could it be due to ions in the oxide?" I tried to answer as well as I could, but it was clear to me that most people in the audience did not believe that I had observed quantum mechanical tunneling. This became a challenge for me, because I believed I had, and for several months afterwards I tried to invent a way to show, by experiment, that I was correct.

One idea I had was to try and measure tunneling at high voltage. If I made a tunnel junction with a very thin upper metal film, or with a film full of small holes, the tunneling electrons should escape from the junction by going over the work function in the top film and into the air. I worked on this for a few months and had achieved what I believed was a modest success, but the experiment was difficult to repeat. So I was not quite ready to publish when, to my great disappointment, somebody named Carven Mead published a short notice in *Proceedings of the IRE* about *my* experiment. It was disappointing to be scooped at least in the theoretical part. I had wanted to show it experimentally as well theoretically before I published, but Mead's paper took the wind out of my sails.

In the meantime I was working part-time on my master's degree in physics at RPI. To finish in a reasonable time I took courses at night which were given to mainly GE people. I also took examinations in subjects I thought I already knew; I remember I took an examination in Advanced Mechanics and got an A. But RPI would not recognize the grade because I had not paid for the course before I took the examination. I was furious, but I was dependent on the system so I had to retake the examination the next year. Then I paid first! I had other problems at RPI, including being unable to make appointments with several professors. Often when I showed up for an appointment, they said they didn't have time to talk then and that I should come back tomorrow. It took me roughly an hour to drive back and forth to RPI from GE, so that added to my frustration. At the time I swore that I would never set foot on campus again if I graduated. I did graduate in 1964, but did not keep the promise as I became a professor at RPI in 1988!

I took a solid state physics course with Hill Huntington, a professor at RPI, who eventually became my thesis advisor. When he came to the topic of superconductivity, he told us that the resistance of certain metals, like tin, became zero below a certain transition temperature. I, of course, did not believe that, because how can you measure zero? But then he said there is a new theory of superconductivity by Bardeen, Cooper, and Schrieffer which states that when a metal becomes superconducting, a forbidden energy gap for the electrons opens up at the Fermi level. This means that there is a small range of energies at the Fermi level that is forbidden to the electrons. Right then I thought I could measure that gap by using my tunneling experiment. So my hand shot up and I asked: "How big is the gap?" But Professor Huntington did not know. I raced back to the laboratory and did what I often did; I asked my friend Walter Harrison. He said he did not know either, but he will look into it and a day or so later he came to me and said it must be a few millielectronvolts. I then told John Fischer about my idea. He thought it was too simple-minded, but being a very positive person he said: "Go ahead and try it anyway." At this time there had never been direct measurements of the gap in various superconductors.

I had never worked with superconductivity and wondered how I could do that. But fortunately I worked at a big laboratory with a lot of resources and many smart people who were willing to help. As a matter of fact most people were willing to give you a whole day of their time. I ended up asking Jake, a coworker, who had a rig suitable for use in experiments in superconductivity. He told me that you had to use a liquid helium dewar which is shielded by sitting inside a liquid nitrogen dewar in order to get very low temperatures. The temperature in the boiling helium is about 4.2 K (about 300 degrees Celsius below room temperature or 4 degrees above absolute zero) while you have 77 K in the outside dewar of liquid nitrogen. Since the superconducting transition temperature of lead is 7.2 K, I chose to make an aluminum–aluminum oxide–lead junction. After having been supervised transferring both liquid helium and liquid nitrogen I was left alone with the rig. I first arranged the measuring circuit to measure the resistance of the lead–aluminum oxide–aluminum junction. It consisted as usual of a battery, a voltmeter, an ammeter, and a variable resistance to be able to apply different voltages to the junction. The first tunnel junction I made did not work, as I had made the oxide too thick, but the second one did! I first measured the junction at room temperature, and then I let it hang from thin copper wires above the helium bath and measured again. And finally I measured it after it was dipped in the helium. There was a distinct change in the current-voltage characteristic as soon as lead became superconducting just as I had predicted. It worked! Many people at the laboratory became very excited, in particular Dr. Charles Bean. He went up and down the long green corridors at GE and explained my experiments to anybody who would listen. He thought that tunneling could reveal the density of states in any metal, which would have been very important, but unfortunately, it turned out that it did not. Fortunately for metals in the superconducting state, tunneling both showed the predicted energy gap and the density of states. Why that is so is not completely clear. But this was important enough because it verified the BCS Theory of Superconductors (Bardeen–Cooper–Schrieffer) by measuring the size of the energy gap predicted in the theory. Superconductivity was discovered in 1911 and had been a mystery until the BCS theory was published in 1957. Since many theorists had their own pet superconductivity theory, at first the BCS theory had difficulties getting accepted. But my experiment really proved that the theory was correct. I think that John Bardeen put in a good word for me because of this, and if I had done the experiment a few years later, even though tunneling is important in itself, I probably would not have received the Nobel Prize.

Very soon I became aware of the fact that professors D. M. Ginsberg and M. Tinkham at the University of Illinois had already measured the energy gap in superconductors using a different technique, and they had obtained a slightly different result. They did it with infrared transmission through superconducting films and could deduce the size of the energy gap. So I asked Dr. Bean about that, and he said: "Don't worry; your method is much better that what they used. They have to worry about you, not the other way around." That was the first time I felt like a physicist: I had done something important.

Guri is born

About this time Inger and I expected our third child and my mother came from Norway to help out. We had purchased a house in Niskayuna so we had plenty of space now, for both visitors and the family. I picked up my mother at Idlewild airport in the trusty Volkswagen and took her home. I was very happy to have her here to help out with the kids and with Inger

My mother holding newborn Guri, August 1960

who was visibly pregnant. Again when Inger, who timed her cramps, decided to go to the hospital, I drove and left her there. This time she was there a long time before they called and said: "Congratulations your wife has got a beautiful girl." So both my mother, John, Anne, and I drove to the hospital to see the new baby and visit Inger. Inger looked tired, happy and beautiful, and my mother could hold the baby. The family was growing. The next day we visited again and could see four or five babies on "display" behind a window. My mother recognized my new daughter right away, but I frankly could not tell them apart. But then we asked a nurse to show us the correct baby, and my mother had been wrong! My mother refused to believe it and finally the nurse had to show the baby's wrist band to convince my mother. My mother's name was Gudrun, but she liked to be called Guri which is a common Norwegian girl's name. So to honor my mother we decided to call the new girl Guri. One result of this name was that she later was contacted by the military for enlistment as they did not recognize it was feminine name!

Since my mother was visiting, when I drove to Toronto to attend the superconductivity meeting, she came along. I had not been invited to give a talk, but I remember several drawings of my experiment on papers and blackboards around the venue. I remember sitting in on a discussion between Conyers Herring and Phillip Anderson where Herring thought the BCS theory was too complicated and Anderson thought it was very straightforward. After that I left, and since I had lived in Peterborough in Ontario, I drove my mother up there so she could see where we had lived. I also showed her the famous lift lock in Peterborough, as far as I know, the highest lift lock in the world. A lift lock is a contraption where a boat enters a big tub of water, and then the tub is raised or lowered by mechanical means.

Talks about tunneling

John Fisher was the first to talk publicly about the experiment in a meeting in California, after I had published the result in *Physical Review Letters*. It was my first sole author publication. Both John and Charles Bean could easily have put their name on the publication, and I would not have objected, but they generously said it was my idea, so I should publish it alone. Since I was working for GE I had to apply for a patent before I published. This took some time as a patent requires that an invention have a practical application. Eventually I came up with something I thought would work. I ended up getting a patent for putting two metals very close together without touching, ranging I believe, from one angstrom to one centimeter. The publication in *Phys. Rev. Letters* was only one page long and referenced only one published paper, i.e. the BCS theory.

I remember that the first time I gave a talk about my superconductivity tunneling experiment was at the University of Pennsylvania, having been invited by Professor Eli Burnstein. I flew down to Philadelphia, and stayed overnight in a hotel. When I woke up the next morning the city was snowed in and nothing was moving on the street. Even the subway was closed. Professor Burnstein called me and said I should wait before trying to get to the university because it was closed for the day. He offered to pick me up later if possible. This he did eventually, and I gave my talk to five people who happened to be at the University for other reasons. It went very well. Burnstein was very excited about the experiment, and I believe that he was later one of my big supporters and helped to nominate me for the Nobel Prize.

I believe the second talk I gave was at the University of Illinois, and I flew there as well, but had to transit in Chicago. It was a very foggy day, and the plane could not land in Urbana. So we went back to Chicago and I decided to stay overnight because they had rescheduled my talk for the next day. When I got there the next day, one of the persons I talked to was Professor John Bardeen. I was anxious because I knew he was very famous and had received a Nobel Prize for being the co-inventor of the transistor. I told him something about the experiment and he nodded his head. He looked somewhat bored, so I stood up ready to leave, but he sat me down again and wanted to know more. My experiment was a beautiful verification of the Bardeen, Cooper, and Schrieffer theory of superconductivity, which at the time was somewhat controversial. Many theorists had their own version of a superconductivity theory and were reluctant to acknowledge BCS.

I recognized very early on that if you tunneled between two superconductors with different energy gaps, you would get a negative resistance. That is an active device and now you can in principle amplify electrical signals much like with a transistor. There was a physics meeting in Toronto that summer and some famous people from Cambridge University, among others Professor Brian Pippard, visited the laboratory. My manager wanted me to talk to them. I was reluctant because I thought they would ask me about what would happen if you arrange tunneling between two superconductors, but they never did! I had not done the experiment yet, but I was sure you would get a negative resistance, and therefore an active device. Since none of them asked about this I was relieved and declared when I got home: "I am the smartest man I know!" For some reason, probably to boost my spirit to do better research at the laboratory, this became sort of a mantra, but the kids took it as bragging and did not like it much. Of course it was not meant literally; at the time there were several hundred scientists at the laboratory all with a better education than I had had and maybe half of them were smarter than me.

However, I managed to do the two superconductor experiment very quickly using the same aluminum–aluminum oxide–lead tunnel junction. Lead has a superconducting temperature of 7.2 Kelvin and helium boils at 4.2 Kelvin so that posed no problem. The transition temperature of aluminum was only 1.2 Kelvin, but fortunately I managed to reach that temperature by reducing the pressure of the helium. The first person who reduced the pressure of helium, or so the

Explaining my result to three real physicists and helpful friends: Walt Harrison, Charlie Bean and John Fisher

story goes, was a technician in Kamerlingh Onnes's laboratory in the Netherlands. Professor Onnes received a Nobel Prize for discovering superconductivity in mercury in 1911. His technician noticed that at 2.2 degrees helium stops boiling; it becomes a superfluid. But when he told Professor Onnes about it, Onnes did not believe him and so refused to even look! It was left to Kapitsa to discover superfluidity in 1937. Anyway, I could get down to about 1.0 K without too much trouble, and I observed the predicted negative resistance by tunneling between aluminum and lead. That you have a negative resistance is very important because now you have an active device. This was long before the integrated circuit was invented. At the time a transistor was as big as a regular sugar cube and eventually IBM started a large program trying to make a superconducting computer using films of niobium and tantalum, metals which are superconducting. Unfortunately both for them and me, they never succeeded. This time when I published the results from tunneling between two superconductors I had competitors: three people from the Arthur D. Little Corporation, James Nicol, Sidney Shapiro and Paul Smith also published similar results. I was surprised that they had done the experiment so soon. There was a rumor that a GE technician had visited Arthur D. Little Corporation and spilled the beans on an earlier date, but I really don't know if that rumor is true.

Friends at Laboratory

CHAPTER 10

I had many friends at the laboratory and my manager Roland Schmitt was very nice to my family and me. He clearly took a big risk hiring me into the Physical Science Branch because I had absolutely no prior track record in research. He had just bought an old cottage at Lake George. Lake George is a beautiful pristine lake in upstate New York which is a very popular vacation place. Roland's cottage was located at Tongue Mountain and only accessible by boat. So he let my family use both his boat and cottage in the fall when he no longer used it. The boat was an old, but beautiful boat with an inboard engine and a rudder, so it took some skill to make it work, but I managed. In addition to the cottage there was a boathouse with several additional beds and a working player piano. The beds were laid out as if in a hospital, and it was clear to me that the boathouse and the cottage had likely been used as an illegal abortion clinic, because it was hidden and difficult to get to. At that time (1960) abortions were illegal and regulated by the states; it was well before Roe vs. Wade in 1973. Roland had many rules that he wanted us to obey. The main rule was that you should use a stick or cane in front of you when you walked because there were rattlesnakes on the mountain. Another rule was that you should never jump or dive into the lake from the boathouse's roof. So we went by the rules. But one day when we were there he visited us with his family. His children dove from the boathouse roof and walked in the forest with no protective canes. So in my family, "Roland's rule" came to mean a rule that could be broken. He soon bought a small sailboat to have at the cottage, and I taught some of his family to sail. Roland had first bought an instruction booklet, and I got a kick out of the first instruction: "Untie the boat from the dock!"

Very many other scientists at the research laboratory played a crucial part in my life. First of all there were the other members of the Physical Science Branch including Charles Bean, John Fischer, Bob Fleischer and Howard Hart. In my opinion this branch was the most scientifically oriented branch at the laboratory. Most of us ate lunch together at the same table in the cafeteria. I always looked forward to lunch, not because I was especially hungry, but because of the fascinating scientific discussions that took place every lunch hour. These were the best "lectures" I have had the opportunity to listen to, ranging over all scientific topics. We also sometimes talked about political subjects and I was puzzled how little respect President Eisenhower received. I have since learned that it is okay to criticize presidents in the USA; very often they are called stupid or even idiots. Since I have had the opportunity to see a few presidents up close, I know that they are all intelligent and 'people smart'. It is grueling to run for president in the USA, and if you are lucky enough to win, then the real difficulties begin. One very active participant at the lunch table was of course John Fisher who was my mentor when I started, and the one who suggested that I pursue thin films and tunneling. I have never asked him if he knew about Leo Esaki who had already discovered tunneling in heavily doped semiconductors, but I suspect he did. Dr. Charles Bean was another member of this group, and he is the most knowledgeable person I have ever met. I was particularly surprised that Charlie knew more about Norwegian culture than I will ever acquire; Ibsen's plays, Munch's paintings or Grieg's music, he knew it all. I have never been very culturally oriented, but after all, I had spent my first 25 years in Norway! He also taught me a lot about research and about life in general. I am proud to say that aside from Inger, he became my best friend. He once told me that he and most people could be married to anybody, which I wholeheartedly agreed with. I told Inger this when I got home, but she was not as enthusiastic about this revelation as I was! Charlie was really responsible for the scientific success of the Physical Science Branch and co-wrote papers with people like Israel Jacobs, Jim Livingston, and Bob Fleisher. Strangely, I never published a paper with him, but I wrote the obituary in *The New York Times* when he died, much too early, in 1996.

At the lab I continued to explore superconducting tunneling for a few years. Roland Schmitt hired a young undergraduate physicist to help me, Carl Megerle, and we worked very well together. I had published alone, but now Carl and I published one paper in *Physical Review* where we put all our ideas together. We also published a second paper dealing with tunneling at very low temperatures where we got help from Howard Hart, who had a helium-3 refrigerator. This device was capable of cooling the tunnel junction down to a fraction of a degree Kelvin and revealed more experimental information. For example, we discovered a bump in the experimental curve for superconducting lead that made me excited. So far my experiments had merely confirmed the BCS-theory, and now I had found an unforeseen difference. The best thing that can happen to an experimentalist is not to prove a famous theory, but to disprove it! It turned out that the wiggles I saw in the experimental curves were due to the soundwaves (phonons) in lead that cause the metal to become a superconductor in the first place. The theorists saw that right away, so I ended up strengthening the theory even more.

Walter Harrison gave me a lot of help writing my papers as I did not have any experience. He was interested in writing a book about tunneling with me as a co-author, but I had no interest in doing that because I thought writing a book would take too much effort. Even writing this book is difficult for me as it is not a "told to" book: I am doing the writing myself. GE hired a young theorist about that time, Charles Duke, who was very industrious, and he ended up writing a book on tunneling instead.

I was still studying at RPI for my doctorate in 1961 while all this was going on. But GE wanted me to focus on the experiment and to withdraw from taking classes, so I did. Dr. Bean said jokingly that studying at RPI interfered with my education! It was not really a joke because I was learning more doing experiments at GE at the time than attending lectures at RPI.

Josephson effect

Brian Josephson had just published his new theory of tunneling, and he participated in a conference GE arranged at Colgate University in 1963. This was the first time I met him, and he was interested in discussing his new theory with John Bardeen. Since I already knew Professor Bardeen a little, I introduced Josephson to Bardeen, and listened to their conversation. Bardeen had already carefully considered the Josephson Effect and decided it was wrong. So after every argument by Brian, Bardeen shook his head and said: "I do not think so" and he finally walked away. Josephson was quite upset and wanted to know if this was the Bardeen of the BCS theory. When I confirmed that he said: "He does not seem so smart." I remembered that Bardeen opened the Colgate conference and said: "I hope to see you all on the golf course, the weather is supposed to be good." My manager Roland Schmitt, who had organized the conference, was not too happy about that remark. This was also the first conference where flux trapping was discussed; i.e. the beginning of the type II superconducting era. Bardeen later apologized to Josephson at a low temperature conference in London for his failure to recognize Josephson's theory.

By 1964, I started worrying that I would be stuck doing tunneling experiments the rest of my life, but John Fisher set me at ease as he often had before. He said that most people were happy to repeat their thesis work all through their career, but I should considering changing fields. I gave a lot of talks at this time and felt I had been lucky to become known so quickly after I entered the field of physics. It became time for me to re-enter RPI, but I had lost my enthusiasm and was reluctant to do so. But now Inger spoke up; she really wanted me to finish my doctorate since she had supported me by letting me study for so long without disturbing me. I recognized that she was right.

I had the chance to visit Cambridge in 1964 and I became very impressed by Brian. We met in a room full of students and professors, and when discussions came to some sticky points, they always appealed to Josephson and said: "Isn't that right Brian?". They did not mean Professor Brian Pippard but Brian Josephson. Coming back to GE I encouraged Milan Fiske to start trying to detect the Josephson effect, I was tempted to do that myself but at Bell Telephone laboratory they had already detected the so called DC effect. Milan had continued to explore the DC effect and the characteristic Frauenhofer patterns, but he started also to look for the AC effect using frequency detectors. If you apply a DC voltage across a Josephson junction it will radiate at at a characteristic frequency. He told me the reason no one had detected the AC effect was that the impedance match between the junction and free space was very bad. I then got the idea to use just tunnel junctions as both a generator and a detector. Diem and Martin at Bell Laboratory had shown in a recent *Phys. Rev. Letter* paper that tunnel junctions could be used as a detector for high frequency. So all I did was put two overlapping tunnel junctions together, such that one served as a generator and the other as a detector. That worked right away and I talked about this experiment when I recieved my Buckley Prize.

American citizenship

Foreign scientists that were hired at the GE research laboratory had to sign a statement that they would become American citizens at the first opportunity, but since I had already worked for GE they forgot to ask me. I could have filed for citizenship already in 1961, but Inger did not want to sign that she would bear arms to defend the USA, so we postponed the decision. The personnel department at GE asked me nicely every year, and in 1964 I gave in even though Inger did not want to join me. It is rather easy to become a citizen. You have to learn some American history and politics, who your congressman is, how many people are in the Congress, etc. You also have to prove that you could write English. When I took the test I had a regular interview, and for some reason we ended up talking about Iceland. So the guy asked me to write down: "Iceland is a gloomy place." When I was finished, there was an Italian family in the waiting room and they wanted to know what I had to write? I tried to tell them that they would never get the same questions, but to no avail, so I told them: "Iceland is a gloomy place." Right away they asked, "How do you spell gloomy?"

But I also had to answer several standard questions to obtain American Citizenship, such as: "Are you an idiot?", "Do you believe in communism?", "Is your main form of income gambling?", and "Have you ever committed adultery?" The answers they were looking for were obviously "no". When I got home I said to Inger that I had won a new freedom today because the US Government no longer cared if I committed adultery. So Inger said: "I think I will apply for citizenship, too!" We both became citizens in 1964.

At the actual ceremony we had to meet at the courthouse and swear in front of a federal judge. There were about 20 of us. We met at 9 in the morning. Nothing really happened for a long time, but at around 10 a troop of boy scouts who had been invited to witness the ceremony appeared. I thought now we will finally begin, but again nothing moved.

Then two unfortunate boy scouts arrived in wheelchairs and finally the judge gave a little speech where he said that as new citizens we would get all the rights and privileges he had. We all had learned in the course we took that none of us could become president of the USA, but I did not think that was very important to any of us. Finally the judge said: "Please raise your right arm and repeat after me:

> *"I hereby declare, on oath, that I absolutely and entirely renounce and abjure all allegiance and fidelity to any foreign prince, potentate, state, or sovereignty, of whom or which I have heretofore been a subject or citizen; that I will support and defend the Constitution and laws of the United States of America against all enemies, foreign and domestic; that I will bear true faith and allegiance to the same; that I will bear arms on behalf of the United States when required by the law; that I will perform noncombatant service in the Armed Forces of the United States when required by the law; that I will perform work of national importance under civilian direction when required by the law; and that I take this obligation freely, without any mental reservation or purpose of evasion; so help me God."*

"You can take your hand down now." And we all said in unison: "You can take your hand down now."

Finishing my Ph.D

I had been tempted to drop my Ph.D. degree, but Inger had put her foot down so I had to choose a thesis. At RPI and at most colleges in the USA, the thesis work has to be done on the campus, thus I could not include the tunneling work I had done at GE. I chose to do my thesis with Professor Hill Huntington. Professor Huntington was a theorist in the Physics Department at RPI and he suggested the problem of scattering electrons in both ordered and disordered copper–gold alloys. He had the feeling that the scattering time would be anisotropic. Since this was a theoretical thesis, I could do most of it at GE. I had already passed the qualifying exam at RPI, and I also had to pass two language tests. I tried to use Swedish and French, but they would not let me do Swedish, so I did German. I had difficulties with German at first, but learned to translate it first into Norwegian and then into English. From French I could translate it directly into English with no difficulty. After my intensive schooling on GE's A, B and C courses, I had little difficulty with the qualifying exam. But then I did not pass because an old rule said I had to take a course in thermodynamics *before* I tried to take the exam. Fortunately, I was the only student who passed cleanly, and Professor Lichtenstein helped me by arguing that the requirement for the thermodynamics course should be waived, and for once, RPI bent their rules.

Walter Harrison helped me with my thesis and was in reality my real thesis adviser. He told me about a new technique developed by Volker Heine in Cambridge, which I used for the CuAu calculation. Now I used a digital computer, which was very convenient, because when the program runs you can change the parameters, like the amount of anisotropy in the scattering time. It is very tedious to write a thesis; there are lots of rules about the size of the margins, equations, and other general appearances. Fortunately for me, I could borrow one of the super typists available to me at GE, so that part was easy. It turned out that Huntington's hunch about the scattering time was not correct, and therefore I never got around to publishing this work. I tried to avoid meeting Professor Huntington after that, because he always asked me when I would write it all up for publication.

My parents' visit

I was short of time however, because my parents were visiting from Norway and I thought it would be nice if they could participate in the actual graduation ceremony. My parents chose to go by boat from Norway, so I sold my well-used but loved Volkswagen Beetle and bought a Chevrolet station wagon so I could meet them at the dock. By now the "Stavangerfjord" boat Inger and I had taken had long since been retired, and my parents arrived on a new boat called "Oslofjord" in the spring of 1964. It was nice to have my parents visiting; I could show them that I had been successful and that our emigration to Canada and later to the USA had turned out well. We tried to be good hosts and travelled around the neighborhood with them. One day we went to Lake George and I rented a boat. We had a nice picnic on an island in the Narrows where my family had camped several times. I also took them on a trip to Washington DC, where my father knew a man in the Norwegian Embassy. This was an interesting trip, because I brought my son John as well. We visited many of the things people do on their first trip to Washington: the Washington Monument was one of them. It was quite windy outside when we visited, and while we waited at the top of the monument for the elevator a man said while rocking back and forth: "In a wind like this, the monument swings a foot back and forth. Can't you feel it?" We could all feel it then, but when I checked it out in the brochure later, the real answer was that it swayed only a fraction of an inch. There was no parking, but a few cars were parked on the lawn outside the monument, so I did that, too, so that my parents did not have to walk too far. Coming back down of course I had gotten a ticket for $35. But the ticket stated that if I did not pay, the price would double. I had no plans to drive to Washington again, so it was a good bet not to pay.

On the way back to Niskayuna, we visited the World's Fair in New York City, and looked at several exhibitions as well as took many of the rides. I remember General Electric demonstrated the nuclear fusion that takes place in the sun, and the demonstration was very successful. A person who I had interacted with at the laboratory, Dr. Henry Hurwitz, was responsible for that display. He was a smart guy, because as soon as the World's Fair was over, he withdrew GE from further fusion research. Nuclear energy here on Earth relies on nuclear fission, i.e. atoms are split into two, the same process that powered the nuclear explosions during

World War II. The sun gets its energy by the opposite process: by fusing two heavy hydrogen (deuterium) atoms together to form one helium atom. Fusion on Earth, including "cold fusion", has been an unattainable dream since 1945.

As for the actual graduation at RPI in May 1964, my parents sat in the audience, and saw their son receive a Ph.D. degree, which made them proud; remember none of my parents had even gone to high school. I was proud as well and thanked Inger for making me stick to the original plan. I had rented a robe for the occasion, and we have some movies of me and my mother posing in our driveway.

My parents also had the chance to visit my mother's friend Rolf Mellerud who had met us when we first arrived on the boat in 1952 about 12 years earlier. He was a real estate broker in Schenectady and also ran a small gift shop. His wife was a Norwegian as well and they used to stop over at our house after church for cakes and coffee. They became substitute grandparents to our children.

A little time after I had taken my parents to JFK for a flight to Oslo, I got a big surprise which I could not believe. I received the Oliver Buckley Prize from the American Physical Society! To me it was completely unexpected, and I could not believe that I deserved it. I was really kind of embarrassed, but of course also delighted. Inger and I went to the March Meeting of the American Physical Society which I believe was held in New York City in 1965 to receive the prize.

Trip to Sweden

In 1965 I was invited on a two-week trip to Sweden together with 19 other American engineers, all expenses paid by the Swedes. I don't think the Swedes knew that I was originally from Norway. I checked with the laboratory and they said okay. It looked like a fun trip to me, starting in Stockholm and ending up in Kiruna, north of the polar circle. But I also had to check with Inger, who was visibly pregnant at the time with Trine, our youngest child.

The expected birth would be one week after I was to return from the trip. Knowing that I really wanted to go, Inger said okay; she is a great woman! To try to save her some work, I tried to get John to come with me, but he didn't want to. To my big surprise, my oldest daughter Anne, who was 7 years old, said she wanted to go and I agreed. In Stockholm she stayed with my brother's family and later in Norway she stayed with my parents.

The Swedes really went all out to give us a good time, and first took us out to "Operakjelleren", an excellent restaurant in Stockholm. Most of the Americans ordered reindeer meat served with new potatoes. The next day I was asked by many of them why the potatoes were so small in Sweden. I had to explain that in Scandinavia it is a special treat to get new small potatoes in early summer, and the smaller the better. Most of the people I was traveling with had difficulties sleeping at night, because it never really gets dark in Sweden in June. We visited several Swedish industries as we travelled north, ending up in the iron mines in Kiruna. In Kiruna they flew two of us at a time in two small helicopters into a hut in the mountain to see the midnight sun. We were served reindeer steak which no one was very happy about as we had eaten it a lot of times by now and the novelty had worn off. The weather did not cooperate so we never saw the midnight sun, but the helicopter ride was very memorable. I had a chance to hike a little on the mountain and came across a reindeer horn, which I took home as a souvenir. It still hangs above my fireplace.

Dotta's cottage

I inherited a cottage in Norway at the same time as my trip to Sweden. My father had three aunts who lived together in Oslo and had never married. Only the oldest of them, Dotta, was still alive. She had one glass eye, had basically lost sight in the other eye and could not hear anything either. She lived by herself on the fourth floor in an apartment building in Oslo with no elevator. My father's three aunts owned a cottage outside of Oslo where my parents had spent a few days when they were newly married. Now the cottage was in the middle of a development project and my brother John had bought a lot from her which he sold at a good profit. Dotta knew I lived in the USA, and she wanted to give me the cottage so I could have a place in Norway to live. Of course I was happy to accept. My sister Mette arranged a meeting with two other cousins who were required by law to agree. In addition she arranged with a lawyer to establish that Dotta was sane. She was declared sane. After the deal was done I stayed back to talk a little to her. She said that she had a thief living in the apartment who hid things from her, but she said she was not afraid of him because he was kind. Then she told me she had recently been at the hospital, and when the ambulance took her there all the bats in the attic had followed the car. She liked that a lot. And she also told me that the bats had welcomed her back home as well. My conscience bothered me a bit after these revelations because I no longer trusted the sanity tests and so I compensated my sister and my cousins somewhat to make up for that. When she died a few years later, my sister Mette helped me by first selling the cottage which somebody removed from the lot. It had become a nuisance for the people in the neighborhood. We then cleaned up the lot, and sold it at a good price. There was no way I could make use of the property.

After about two weeks I picked Anne up in Norway. She was very happy to see me, but really desperate to go home to her mother; remember she was only seven. It was the first time she had been away by herself, and I told her she had been very brave. I was anxious to get home, too, because when someone asked me how many kids I had, the answer was: "Don't know, either three or four." Turned out it was still only three when Inger met me at the airport. And the first thing Anne said to Inger was in Norwegian: "Se mor, jeg har faatt en klokke! (Look mother, I have gotten a watch)". Since we spoke mainly English at home then, Inger was surprised to hear Anne speak perfect Norwegian.

Trine is born

About a week later, Inger gave birth to our fourth and last child. This time, I had very good help from our next door neighbor, Helen, and could manage it all by myself. I actually sent John to a summer camp for 3 weeks, but he did not want to go. He was twelve years old and quite sensible, so when I explained it would help both Inger and me, we made an agreement. If he went to camp this time, he would never have to go again if he did not want to. He never did! Now the trip to the hospital had become routine, but you are of course anxious that everything goes well. There was no ultrasound so of course we were curious about the sex. Inger's doctor had prided himself on predicting the sex of the baby depending on how it had been carried, and he predicted a boy for us. It was, however, a lovely girl. When I asked the doctor, he said he had predicted a girl and showed me that he had written it down in his notes. It turned out he used this method to predict: he wrote down one sex and predicted the other. This trick made him a perfect predictor.

Our children: Guri, John, Anne and little Trine, fall 1965

Inger's education

A few years after Trine was born, it was my wife's turn to be educated. Like most people her age in Norway, she did not go to high school, so she had to take the high school equivalency exam before she started at the Schenectady community college. After she had finished two years' worth of study there, which probably took four years, she transferred to SUNY, Albany. She and our next door neighbor started together, but Inger was the only one who had the strength and will to finish. And all along she took care of me and the kids as usual. She graduated with a liberal arts degree in 1983 with German as the main subject. I think I have never told her how proud I am of my wife, but hopefully she will read it here? For Inger's graduation her mother came to the USA to participate, and so did our 3 youngest children who by then were young women and lived independently.

Our Daughters, Anne Guri and Trine at Inger's graduation

Inger's mother at Inger's graduation, spring 1983

CHAPTER 11

MY TRIP TO MOSCOW

Tunneling had by now become a popular field and there was at least one session about tunneling at every March meeting of the American Physical Society. Tunneling in superconductors had also become a major subject, and when a low temperature meeting was arranged in Moscow I received a special invitation to attend. Actually several scientists from the GE laboratory attended the meeting, and since this was during the Cold War we all needed special visas. We were required to prepay for our stay, and there were three classes to choose from: economy, normal, and deluxe. So we all applied for normal, but then I was assigned to the deluxe class. Roland, my manager, said I should refuse and go for the standard class, so I wrote them back and got the answer: "Ivar Giaever will go deluxe or not at all". I of course was happy about that, and Roland gave in so I went deluxe.

Before I left on the trip, I was interviewed by a person from the FBI about the trip. I was not exactly asked to spy, but I believed maybe that was the purpose. This was the height of the Cold War and the Soviet Union was regarded as a big threat. A list existed at the time, real or not, specifying which cities the Soviet Union would bomb first in the event of a war. Since Schenectady was *not* on the list, people here were upset. After all we were known as the city that lights and hauls the world! Remember that both General Electric Company and the American Locomotor Company (ALCO) factory had major manufacturing businesses in the city. I had to swear on the Bible that I would tell the truth when I returned home. I told him that I was not religious so swearing on the Bible meant nothing to me. That created a lot of trouble for the FBI agent, and I can no longer remember how we settled this issue. Since then I have learned to swear on the Bible when asked!

I arrived in Moscow together with many other physicists I knew; in particular I remember Allen Heeger and Bob Schrieffer who also brought his wife along. When we arrived there was an immense queue at the airport; it turned out that everybody had to show that their baggage check agreed with the number on the checked suitcase. Since we were not used to this in the US, people went through their pockets to try to find the correct paper documentation.

We all stayed in a gigantic hotel in Moscow. I think my room was on the 8th floor. The first morning there was an incredibly long line for breakfast. To pay for the breakfast we used food stamps which the meeting organizer had sent us before we left. Because I travelled deluxe, I had plenty of stamps and I gave the waiter a couple of extra tickets as a tip. It paid off, because next day I was served very quickly; it seems that tips are appreciated even in communist countries.

There was lot of interest in my talk so I was very happy about that. I also spent a fair amount of time going to other scientific talks. But since I travelled deluxe I had a driver, a car, and an interpreter at my disposal, so I also got to see lots of sights in Moscow. The interpreter was a very nice-looking woman, but also a very straight-laced communist. It became obvious that no flirting was permitted. I remember we went to the space museum in Moscow where I saw a copy of Sputnik and also the satellite that carried the dog Laika into space. Both Sputnik and the satellite looked like they had been put together rather clumsily by a blacksmith. I found this surprising, but as the Soviet Union was leading the space race, they obviously had much more powerful rockets than the US.

My communist interpreter had received permission to go to England on a study tour and she was very excited about her upcoming trip, I tried to tell her about how some things were done in Western countries. When it came to food she asked how this was managed, and I said you pay for it with money of course. She thought for a moment and said: "How inconvenient." Remember we all were issued food stamps before we entered the Soviet Union.

Since I had so many food stamps I took her with me to a fancy restaurant and ordered caviar which was readily available at that time. She ate well and was very proud of the treat. She also showed me some new apartments in Moscow which were kind of small, and I remember that the kitchens were often shared with one's neighbors.

When you ate dinner at the hotel, you had to wait until the table filled up; they would only serve 10 people at a time. This of course was very efficient for the waiters, but not very convenient for the guests. One time I sat down to dinner and everybody ordered beef stroganoff, but since I had eaten it the night before I wanted something else. I went through some twenty items on the menu, but the waitress said they were out of everything I asked, so I finally succumbed and ordered beef stroganoff.

At the hotel we had to wait endlessly for the elevators to take us both up and down. So one day I decided to walk down the 8 floors. When I got down I could not get out of the hallway because the door was locked, and I had to start walking back up. By hammering on the doors at various floors a female guard finally heard me and let me out of the hallway. It turned out that on every floor in the hotel a woman sat guard and the purpose was to prohibit any "improper" behavior. For example,

Bob Schrieffer invited me up to his room once for a drink, and his wife Anna was with us. The female guard insisted that we keep the door open, and only then was I allowed into their room.

I discovered that the young people in Moscow had devised a scheme to avoid the female guards when I took a trip on a hydrofoil boat on the Moscow river. It was a beautiful journey, going through woods of birch trees. I talked with many young people on the trip who were very interested in knowing about the British spy, 007, who for some reason they had heard a lot about. On these boats you could rent cabins with a bed for a few hours. We rented one to gulp down some vodka. The guys were very good at that. I tried to do the same, but two nice girls warned me about getting drunk, so I for once followed their good advice.

Peter Kapitza invited me out to a restaurant for dinner one night, which was very nice. He is the son of the Nobel Prize winner Pyotr Kapitza. He was writing a book about world population and argued that overpopulation would not be a problem. The common view at the time was the reverse — namely, that pretty soon we would be unable to support the growing world population. At the time I did not believe him, but as it has turned out, he was correct. You need 2.1 children per woman to keep the world population constant; today in all developed countries the average number of children per woman is fewer than 2. I think the reason for this is that neither our cars nor our refrigerators can support many kids; and of course "the pill" is very important. We discussed the issue over two bottles of sweet wine. As soon as we sat down before we even saw the menu, he ordered the wine. The wine appeared reasonably quickly. He tasted the wine and said: "I was afraid of that." I asked what the problem was and he said that the wine was sweet. I asked if he could change it, but that was not possible. Apparently in a restaurant in Moscow you had to be happy with what you were served.

I visited a low temperature lab at the University of Moscow and it was more or less similar to the one we had at GE. They were in the process of repeating my experiment and were very proud of a new evaporator they had just acquired. It had no trap for liquid nitrogen however, which was necessary at the time in order to get a low enough pressure to evaporate metal, but they hoped it would work anyway.

When I got back to Denmark I stopped by a bar at Copenhagen airport where they had a sign: "Have a drink, we take any currency!" Turns out they did not take rubles when I tried to get rid of mine. The bartender said the rubles were really not a currency, and I let it go at that. When I got to Kennedy Airport I was struck by the fact that everything was straight and looked so elegant. In Moscow everything was a little out of line, as though things were built by amateurs and not by skilled people. At GE I was interviewed by the FBI again. This time I swore on the Bible and the agent looked very satisfied.

MORE RESEARCH

12

The DC transformer

I also had a nice trip to Brighton in England in the summer of 1965 when there was a small conference on superconductivity. I remember going to the beach and was surprised that everyone was sitting down in chairs and no one was playing on the beach. When I got my swimming trunks on I discovered why; the beach was full of small stones, just the right size to make them too painful to stand on. I remember I crawled on all fours to get into the water. People mostly talked about type II superconductors at this conference. They claimed that the voltage you get when you apply a current to a type II superconductor is due to the fluxons moving across the conductor. That sounded strange to me, so I decided to test it by doing the following experiment when I got back to GE. I first evaporated a film of tin onto a glass slide and on top of this tin film, an insulating film of silicon oxide. Then to complete the structure, I evaporated another film of tin on top of the insulating silicon oxide. By carefully applying a current to the bottom film, the fluxons should start to move, and as a result cause a voltage to appear along the film. But since the tin film on top of the insulating silicon oxide is very close in proximity, the same fluxons will also cause a voltage to appear in this film. This is exactly what happened when I did the experiment. Similarly, when I evaporated two parallel films on top and put them in series, I expected to get twice the voltage, i.e. making a DC transformer. I was quite proud of this invention even though it had no practical value. I got a call from the person dealing with the elevators in the Empire State Building who had read the press releases and wanted to try it out. He was quite disappointed when I told him he would need a dewar roughly the size of the Empire State Building to use the invention!

Some other experiments

Werner Kanzig was one of the first scientists I met at GE. I remember the first time I talked to him I happened to be filling a dewar with liquid nitrogen from a tank, right after he had finished filling his dewar. Without warning he lifted his dewar and poured liquid nitrogen on my hand. I was very puzzled but then he said: "You will become a good experimentalist because you did not drop your dewar." He decided to go back to Switzerland and become a professor. One of his first students was Hans-Rudi Zeller, and GE asked me to go to Switzerland to interview him as a potential employee and coworker at the laboratory. The day I arrived in Zurich he was to meet me at the airport, but he never showed up. When I talked to him later, he said he had been there but did not recognize me. This was not a good sign. The next day he celebrated his Ph.D. degree by inviting his fellow students to a feast in a forest. I guess he had my visit in mind when he selected the date; that was a good sign. He had arranged for somebody to grill a piglet on a spear up in the mountains outside Zurich. It was a beautiful night: very clear, and lots of stars were visible because we were so high up. We could also clearly see a satellite that Werner pointed out. I guess he had prepared for it and knew the time it would be overhead. I was charmed by this event, and since I liked Hans-Rudi, I hired him to work with me as a postdoctoral fellow.

We worked very well together. One problem we tackled was to investigate the size of superconducting particles. The specific question being asked was, how small can a particle be and still be able to enter the superconducting state? To do this experiment we first evaporated an aluminum film, and then oxidized it for a short time. Next we evaporated tin in a thin layer such that particles would form, waited for the tin to oxidize and for the aluminum oxide to grow thicker, and finally finished the sandwich by evaporating aluminum on the top. When we applied a voltage to this structure, we expected a nonlinear current–voltage characteristic to appear. The reason for that was that the electrons had to flow through the tin particles to get from one aluminum film to the other. To get an extra electron onto a small particle, it had to be charged up, and that required energy. The curve really looked like a giant Kondo effect, a tunneling experiment I had never really believed in. Now, if the tin particle becomes superconducting by cooling the structure down, we expected the energy gap in the tin particle to cause the current–voltage characteristic to be even more nonlinear. By applying a large magnetic field you could quench the superconductivity, and the current–voltage characteristic would change back to what it was. We were never able to find the limit of the particle size, because the magnetic field available to us was not large enough to destroy the superconductivity in particles below 30 angstrom. So we concluded that tiny tin particles were still superconducting in the 30 angstrom range.

We also worked on a problem where we saw regular periodic bumps in the curves using two superconductors. This was a curious problem which I never could understand completely, but we got a couple of papers out of our observations. We also tunneled through evaporated semiconducting films. By using cadmium sulfides and shining light on the tunnel junction, we could change the resistance to zero if both the metals were superconducting. It seemed likely to me that this would eventually have practical applications, but at the time the low temperature required was a problem. But we had an interesting time.

SABBATICAL LEAVE IN ENGLAND

CHAPTER 13

In 1968 after having worked at the GE Lab for over 10 years, I decided to apply for a sabbatical leave which GE offered. The unwritten rule was that if you could get some support from a foundation, GE would then pay the rest. I decided to apply for a Guggenheim fellowship and they gave me a stipend. What they said in the agreement was that the money is yours to do with as you please and whatever you do, please do *not* wrwite and tell us what you did! GE agreed, so now we had to decide where we wanted to go. Since we had three children in school and one who was only four years old, we decided on England because there would be few problems with the language. And since I had already met Professor Brian Pippard I wrote him and asked about the possibility of coming on sabbatical leave. He reacted positively to the idea, and since he had become headmaster at a new college, Clare Hall, he also offered me a new, fully furnished house at the college. We decided to take our old Chevrolet station wagon with us and therefore went to England by boat. We settled on the *SS France*, and planned to take a trip in Europe first, a vacation in Norway, and finally a ferry from Norway to England. I remember driving the station wagon down to the docks in New York City to hand it over to the *SS France's* crew. The first thing I noticed was somebody pushing a refrigerator off a truck onto the dock — the refrigerator got crumpled in the bottom. I managed to locate the person who was responsible for loading cars onto the ship, and when I gave him the key he said: "It is customary to give us a tip". I asked him how much and he said: "Oh, about $200." I thought for a moment and decided to give him $20, which he took grudgingly. I was nervous that he would damage my car, but when it was unloaded in France, it was okay.

When we finally entered the ship, at first we had no cabins even though we had booked months beforehand. After lots of confusion and discussions we managed to get two cabins in the same corridor; one for the three kids and one for Inger, me, and Trine, our youngest daughter who was 4 at the time. The boat trip was kind of boring to me; I guess I am not suited for cruises. I remember that the bigger kids had a good time and were running all over the boat. They especially liked the movie theater, where they could see R-rated movies. We were very strict at home and never let them see R-rated movies unless we were with them. I believe they all saw "Midnight Cowboy" on the trip, at least I remember that Inger and I did. I also remember a movie in French "*Les Parapluies de Cherbourg*," which I saw, and the language was so simple I understood most of it. I was proud I had not completely forgotten my French.

My wife entered a ping-pong tournament and won. Truthfully she was not very good in the game, but her opponents were two young French girls and she ended up giving the girls the prizes. I found a person I could play chess with. He flattered me and said I played like Bobby Fischer, but unlike Bobby Fischer, I lost most of the games. We all ate lunch and dinner at a small table suitable for my family only, and at every meal we got two bottles of wine, one red and one white. I had been in the USA for about 15 years by now and had gotten addicted to martinis. Since you had to pay a ridiculous amount for cocktails on the ship we stuck to the wine.

Five days later we arrived in France, and with most of the luggage loaded onto the roof of the car, set out for Europe. We had no reservations and just drove until the afternoon before finding places to stay. I remember one place in particular where we rented two small houses which reminded us of the houses in the "Flintstones" TV series, one of our favorite TV shows. Another place we also enjoyed was Paris. We got there after dark, driving in a narrow one way street with lots of cars after me that were flashing their lights. It took me some time to realize it was me they were trying to hurry up. In France at the time anyway, the people are notorious for having difficulty understanding foreigners trying to speak French. Since French used to be a major language, the French often do not speak any other languages. So my son John, who was 15 at the time, and I tried to speak French to ask for some information, like: 'Ou est la Place de la Concord?" After having repeated this 4 or 5 times they finally understood it. When we were finished trying to communicate, Anne who was 11 at the time, and who had taken a few years of French, told us what we should have said. I had several scientific friends in Paris and one of them, Pierre de Gennes, invited Inger and me to his home for dinner. We had to feed the kids before we went, and we took them to Wimpy's, a fast food place for dinner. Since we were late Inger and I decided to go back to the hotel and change and the kids could follow, as the restaurant was only one block away. Everything went smoothly, except when John came back, only our younger daughters Guri and Trine were with him. When we asked where Anne was, he said she wanted to go back alone. But when we went out on the street she was nowhere to be seen. Inger and I ran frantically through the streets trying to find Anne. She had a red coat on so she should have been easy to spot, but this was rush hour and the streets were filled with people. We must have spent about an hour searching, before Inger spotted her.

She knew she was lost, and she had walked right by the hotel without recognizing it; even though she could speak some French, she was too shy to ask anyone, or maybe she did not know the name of the hotel. We were very relieved. Then we called de Gennes and told him the story and that it was late, but he said it was OK, so we took a taxi and had a nice French dinner. Being in France lots of alcohol was served and when we were going back to the hotel one of the scientists who had attended offered to drive us home. That was a harrowing experience. He drove very fast through the relatively empty streets, but when the car in front of us stopped for red light he was unable to stop, swerved around the car, and narrowly missed the car in the cross street. His only remark was, "I did not know cars stopped for red lights so late at night in Paris!" Inger and I were quite shaken for the second time that day.

We of course wanted to visit Epernay where I had stayed in my student days. When we got there I was unable to recognize exactly where I had stayed as a student. It was right opposite the big Catholic church and a restaurant where the parishioners used to have a drink after services, but the restaurant was now gone. We stayed at a small hotel nearby which was a house of ill repute when I lived there. Before we left we visited Moet Champaign and were offered a bottle. We shared some of it with the kids, and when we left Epernay they were singing in the back seat, but soon they all passed out.

All in all we had an interesting and educational trip with only a few mishaps. I impressed the children by ordering escargot, which is a snail dish. One rule Inger instituted was that if you order something, you had to eat it. It was a good rule. One time all the kids ordered Beef a la American and it turned out to be Beef Tartar. Inger changed her mind and I got mad at Inger. She insisted on asking the waiter to cook our hamburgers, which they did. We also learned that if a restaurant in France had tablecloths, it would be expensive. Another puzzle to me is that the mustard in France is wonderful wherever you are, so why can't you buy it in the USA?

As for driving in general, people drove much faster in Europe than we were used to. I had a Chevrolet station wagon with lots of power, so sometimes when small cars behind me flashed their lights, I simply floored the gas pedal and left them in the dust. But Inger was not so happy with my driving then. Some other people disliked my driving as well or maybe it was the fact that our huge American car was plastered with flower power stickers? Once in Germany when I tried to pass a slow truck, it speeded up and would not let me back in the right lane. I had to do an emergency brake to get back in the right lane. Then the truck slowed down to walking speed, but I was not tempted to try to pass again. Eventually we ended up in Norway at my parent's cottage on the Oslo fjord that we had visited several times over the years. We spent the rest of the summer there with my parents, and had a very good time.

In the beginning of August we set out for England by taking a ferry from Kristiansand, Norway to Newcastle, England. I was somewhat nervous learning to drive on the left, with the big American car, because the steering wheel was now on the wrong side. The first thing I noticed was that a large number of cars had a big capital L pasted on the rear. The first thing that struck me was that they were awfully nice reminding me to keep to the left, but of course it was only a large number of people who were learning to drive. It turned out that it was quite easy to drive on the left, except if you had to pass a car. Normally one of the older children put their head out of the window, and announced: "It is safe to pass now Dad, if you hurry!" Scary to begin with, but sometimes you really have to pass those slow lorries.

Clare Hall was an architecturally renowned modern apartment complex built for visitors to Cambridge, and it turned out to be a lovely place to live. The complex also had three separate houses and we got one of them. The house we moved into had three bedrooms and a big tree outside in a fenced-in yard. The first thing Inger and I did was to open a bottle of red wine to celebrate our luck with the nice accommodation.

Clare Hall. Our house with a tree in the garden, in front, 1969

I was standing in the hallway outside of the kitchen trying to pull the cork out, and suddenly it popped. It made the bottle move violently and red wine splashed onto the pristine wall. We tried to wash it off, but it did not work, so we scrubbed it a little every day, and it was almost gone when we finally left Cambridge.

There were many visitors just like us at Clare Hall, and therefore it was easy to make friends. We asked Mrs. Pippard if she knew any babysitters, as she had three children herself. She said she knew many, but they do not sit for American children! As it turned out both John (15) and Anne (11) ended up by being very popular babysitters in the complex, and since we normally were home and right next door, people were not too concerned with Anne being only 11 years old.

One of the first things we had to do was go to the grocery store, and it was very different from what we were used to in the US. We parked the station wagon outside the store, and bought anything and everything

Inger and our 4 kids in our yard at the house in Clare Hall

we could think of, filling several of the small baskets they had in the store with groceries. When we got to the cash register the woman looked kind of puzzled but checked everything through the cash register. But then nothing happened. It turned out that you had to bring your own bags into the store, and pack them yourself. After some time they found some big cartons and we managed to take it all home. The problem was exacerbated because few families in England had so many children and so their grocery lists were much smaller than the one for our family of six. It was also more common to go to the store every day.

One of the first orders of business was to register the kids in school, and we quickly learned that when you say "public school" in England, it actually means private school. In any case, because our children spoke 'American', the local schools would not take them and we had to send them to private schools. We took the advice of many people who recommended Perse Schools. Anne registered probably in one grade higher than she should have, and was very busy with homework the whole time we were in Cambridge. Guri went into the more or less the correct grade. She is however left handed, and in her class they had to write on cheap paper with fountain pens that were dipped into open ink bottles. Guri spilled ink all over her reports. Finally the teacher said that her reports were so messy that she would not correct them. I decided to intervene, and tried to talk the teacher into letting Guri use a ballpoint pen or better yet a pencil, like they did in the USA, instead of an old-fashioned pen. But the answer was NO. I tried a little sarcasm, and asked when had children stopped sharpening feathers for writing reports, but to no avail. I lost that fight. Instead of recess, Guri now had to stay inside and work on becoming right-handed, which she never accomplished.

John started at Perse High School; for him the school was a great success. He had grown tired of school in Niskayuna, and was just coasting. He was interested in sports, but in Niskayuna he was just an average athlete. At Perse however, he became a star athlete and played rugby. Most of the students at Perse were really nerds. In order for him to get to school we had to supply him with a bicycle; actually we bought four used bicycles for the family. The guy who sold them to us called them machines, not cycles. They were very old fashioned, probably like 30 years old. In England people do not throw things away, like we do in the USA. If it still works it is OK. A person wrote in the newspaper to the director of the railroad asking why it took 1 hour and 45 minute to go from Cambridge to London by train, when it took 1 hour and 25 minutes in 1880? The answer he got was brief, but accurate: "Dear Sir, the reason is that we use the same equipment!"

Anyway, we supplied John with a bicycle, and he was happy in school. One day I ran into my son with his bicycle on Newton Bridge in downtown Cambridge in the middle of the day. His face was pale, and I asked him: "What is the matter?" He had played rugby and hurt his hand. The hand was quite swollen and I decided to take him to the hospital where they found out that his hand was actually broken. I was quite upset by this and decided to complain to the school. I got an appointment with the headmaster. The first thing I saw when I entered his office was a tiger skin lying on the floor in front of his desk. It was facing the visitors with the teeth showing. The headmaster was very polite like most Englishmen, and offered me a glass of sherry. They all offer you sherry in England before you can state the purpose of your visit. When I finally got to the point of asking how they could send a boy home on his bike with a broken hand, he said: "They did not know it was broken." The coach had not heard any cracking when he moved John's hand back and forth, and therefore he did not think it was broken! He was not very apologetic, about the incident. Perse School at that time was reported to give unruly students a thrashing with bamboo sticks, but I am not sure if that was just a rumor.

I felt sorry that my son went through this awful experience and to cheer him up we bought him a used moped to make it easier to go to school. He loved that moped; I think it gave him more status among the boys at his school as well as the girls at Perse's sister school.

Our daughters' school was close enough that they could walk to school, but when it rained my wife drove them along with Pippard's children. At Perse School all the children were required to wear uniforms which Anne and Guri did not like. But Inger and I thought it was a very good idea because there was no question in the morning what the children should wear. We never bought the expensive coats however, and were constantly reminded by the school about this aspect. It was hard to ignore Guri as she had an orange ski jacket in lieu of the classic navy blue 'cloak'.

We also had to take Guri to the eye doctor when we were in Cambridge, which was no problem because England had socialized medicine. The doctor's office was very small, so when the doctor performed the eye exam she used several mirrors so that the distance from the eye chart was sufficiently large. It turned out that Guri badly needed

John on his moped at Clare Hall

glasses even though she had been checked at the school in Niskayuna right before we left. The glasses were free, which was nice, but to get the free pair there were only four choices of frames. They were all the same shape, and differed only in color: black, blue, pink, brown and black. To my surprise, Guri chose the black frames. If you paid extra you could get nicer glasses, but against our kids' wishes, we stuck with the freebies as we were always careful with money.

To give Inger some free time we also enrolled Trine in a school for children. Since she was four years old, we had to walk her to school and pick her up. The school was in terrible physical shape and very primitive. The children learned to write on slates with crayons. Trine lost her crayon on the floor one day, and it disappeared into a hole in the floor. She was punished for this and had to stand in the corner for a certain time period. One day when I got close to the school, I heard all the children repeating multiplication tables in unison: "One two is two. Two twos are four, three twos are six, four twos are eight," etc. When Trine and I were walking home hand in hand, she looked up to me and asked: "Dad, what is a two?"

Right after we arrived in Cambridge I contacted the physics department and said I was going to spend a short year here. They asked what I planned to do, and I said: "I plan to study biophysics". The answer I got was "You poor man, there is no such thing". That was news to me, but in many ways they were correct. I have worked in biophysics now for more than 30 years, and normally when I meet other people in biophysics, it is very seldom that their work has anything to do with my work. A practical definition might be a physicist that likes to work in biology.

At Clare Hall I ate lunch at the cafeteria most days, as you could eat with other professors, and lunch was free. It was nice to have well-prepared lunches, and normally we had interesting discussions. I can remember one where we discussed Latin, and whether Latin should be taught to school children. Since I am a practical person I argued that Spanish or French would be better, while most of the people present believed you should teach Latin. After the discussion had gone on for some time, an old professor sitting at the end of the table interrupted and said: "I agree with the young man, I do not think you should teach Latin, I think ancient Greek would be much better!' Once I came to the lunch table all excited because I had read that somebody had discovered quarks, and I announced it at the lunch table. A famous theorist asked: "Where did it happen?" and I said: "In Australia." He said: "If quarks are ever found it won't be in Australia." I was taken aback by this snobbish remark, but I have later realized that it is true. To do good science today, particularly in high-energy physics experiments, you need good equipment, groups of excellent people, and facilities, all of which are limited by funding. The days of a single scientist making a ground-breaking discovery by working on his or her own are long gone. I was fortunate that in my early career I could work by myself. At that time it was common to have one or two authors on a paper. Today it is often 6–10 authors even on rather ordinary papers. The reason is both the pressure to publish and because people tend to work in larger groups as then have a greater chance of getting funded.

There was also a dinner at Clare College once a week where you could participate with your wife and children. But Professor Pippard's rule was that the children had to be able to contribute to the conversation. So with the children's consent, Inger and I normally went to the dinners alone. The Clare Hall College had a good cook who was willing to experiment with the dinners. We were asked for Norwegian recipes and suggested a common Norwegian dish, "farikal", which is sheep with lots of cabbage. We insisted that she should use lots of cabbage, but she fell short. Most people who are not familiar with Norwegian food do that. Once she served sheep brains at the weekly dinner, which I found delicious, but when Inger found out what we were eating, she felt sick. Scrapie's is a disease that affects sheep's brains, related to bovine spongiform encephalopathy (BSE) or "mad cow disease", a variant of Creutzfeldt–Jakob disease, so I doubt people eat sheep brains anymore.

After a month or so in England, I went back to the USA to attend a conference on superconductivity at Stanford University. I do not remember too much from this conference except I got drunk with my friend Walter Harrison who was now a professor in the applied physics department at Stanford, where Walter graciously took care of me. This was the conference, I believe, where Bill Little presented his theory of High Temperature Superconductors and reportedly had sold the patent to somebody. This made Phil Anderson very angry, and he said Bill should be sued for fraud. Phil had no faith in that theory. Anyway, it was a nice conference; I was away for about a week. When I returned to Cambridge, Inger met me in a miniskirt. At the time miniskirts were quite new. They really started in England because children's clothing was subsidized by the government, and bigger girls found that they could cash in on the subsidy by wearing short skirts. Long-legged Inger looked a little unsure of herself in the outfit, but to me she was beautiful and sexy.

Back in Cambridge I got a letter from my present English bank asking me to come down for a chat with the manager. When I left Schenectady I had arranged for GE to deposit a certain amount of money in my account every week. I met the manager in his office, and he first served the traditional sherry. After some polite phrases he cleared his throat and said: "Could you please put a little more money into your account, we have carried you for nearly two weeks". The arrangement I had made at GE had not worked, and the account was badly overdrawn. I had used my American bank for about 20 years, and once when I accidentally wrote a check for $25 and only had $20 in the checking account the bank fined me $10 the next day. So there are differences between countries!

Sir Neville Mott was the head of Cavendish laboratory when I was in Cambridge, and I ran into him frequently. He asked me one day if I had read his new book, and I was not aware that he had written one. So I said no. He then asked me if I wanted a copy, and I said that would be very nice. So he said come into my office and there he produced a copy and gave it to me while he said: "That will be 27 pounds". I was surprised, and the first thought which came to me was to wonder if he took credit cards, but fortunately I caught myself and simply gave him the money.

Phil Anderson was also in Cambridge part of the time when I was there, and he gave a course in solid state physics, which I took. In England there are in general no graduate courses you have to take for a PhD; a thesis is all that is required. But after Phil had given four or five classes about the basic principles in quantum mechanics dealing mainly with superconductivity and Green functions he asked, "Are there any questions?" and a student raised his hand and said: "Professor Anderson, what is a phonon?" (Phonons are the quantum mechanical representation of sound waves in a solid). I felt sorry for Phil at the time, because the question made clear that this student and maybe the whole class had not understood anything of the previous lectures. Phil patiently tried to explain the concept. After that I tried to persuade Phil to give problems in the course, but he steadfastly refused. Finally in the last lecture he said: "Explore the two-particle Green function and use it to explain superconductivity". So Phil finally gave a problem which I am sure no one in the class could do, including me.

Cambridge of course is a very famous university, but I found it rather poor when it came to equipment. I had wanted to continue to do some tunneling experiments, but decided against it. The reason was that the equipment I had access to at GE was much better than at Cambridge so I just helped various people in the laboratory. This was the time when people at Cambridge "tunneled" through normal metal between two superconductors. I never believed those experiments, but I was in the minority. I rapidly discovered that at scientific discussions in Cambridge people did not seem to care about what was correct or not; the point was to win the argument, and I am good at that. There is really a fundamental difference between scientists in USA and England. When an American comes into your office he normally starts with something like this: "I have worked my ass off on this problem and have not gotten very far, but I have made some progress the last few months that I would like to talk to you about." When an English guy comes in he says something like this: "When I walked in this morning I had an idea that I would like to discuss with you." Then he would start to write on the blackboard and you would know he had been working on the problem for weeks. I think the difference boils down to the fact that Americans take great pride in working hard while the English are more concerned with appearing very smart!

Our family has always liked to do various sports, and in England we ran a lot with the children, timing ourselves on how fast we could run a mile and what not. Right behind our house was a rugby field we liked to use because of the beautiful green grass. I never saw the field used for anything the half year or so we were there, but some guy chased us off. He said that if he let us run on the field, pretty soon the whole of Cambridge would come and the field would wear out. We continued to go anyway as he didn't always show up, and this was well before jogging was popular. We also played a lot of tennis, and it was nice because of the grass courts they had. I had started to learn to play tennis maybe three or four times in my life, but this time it finally stuck. In England they close the courts every Sunday, which is when they are most used in the USA.

One day at breakfast Inger and I noticed a lot of commotion outside the house and when we asked what was going on we were told it was a long walk, maybe 50 kilometers, I cannot quite remember. It was the very famous Oxfam walk. At the time I had never run a marathon and had no idea how long it would take to walk 50 km, but we just decided to participate. We thought we could take it easy, see the countryside and stop to eat or drink at a few pubs. So we rode the bikes to the starting line which was maybe a mile or two from us, and entered in good spirits. I believe the walk started at 9 A.M. It started to rain right after we began walking so there went the pubs. It was a very long walk, a very, very long walk. We had ample chances to get a lift back to the starting line, and I wanted to accept. But Inger said no, she was determined to finish. We were one of the very few who actually finished the whole walk. When we finally got back to the starting line Inger was exhausted. She was in no shape to cycle home, so I took the two bikes and went home for the car to pick her up. But when I got back she was nowhere to be seen. One kind man had taken her down to a basement and given her a shot of cognac, seeing she was completely exhausted. When she appeared, I put her in the car, and when we got home she had to walk through the girls' bedroom before she entered our room. But as soon as she saw a bed, she collapsed in it and stayed there for two days, the cognac was probably also partially responsible. I had blisters covering the soles of both feet, but managed to go to the university the next day, mostly just to brag that we had finished.

As I had decided to learn some biophysics while in Cambridge, I took several courses at the undergraduate level. One course on DNA was given in a big classroom with seating for maybe 200 students, but the course was popular and the room was always completely filled. They had set up a TV in a different place and I normally watched the course from there as it did not feel right to occupy the spaces of the real students. At home I was reading Jim Watson's books; first the popular book, *"The Double Helix"* and then the scientific version *"Molecular Biology of the Genes"*. I liked the scientific book very much because of Watson's very clear writing style. He tells it straight even though he is sometimes wrong but that is OK. Most biology books in the past said: "It may be this way, but it could also be that way or maybe even this way, we are just not sure." I had cocktails in his house at Cold Spring Harbor once; he had a beautiful Scandinavian house which was built inside an outwardly ordinary-looking house. As soon as he had served us a drink he came back and said: "I am going to eat dinner with my family now, so you have to leave." I also took a course in colloidal chemistry at Cambridge. The professor came in with a bottle of colloidal gold and he said that it was really an unstable solution, but since it was prepared by Faraday, it clearly keeps for a long time! Only a chemist will fully appreciate this joke.

We travelled a bit in Europe that year, and we went to Norway at Christmas time. It was always difficult to get across the British Channel, because we never ordered tickets early enough. When we went to Norway at Christmas we had to take the ferry from Dover to Calais, and drive to Norway. Going back we took a ferry from Gothenburg, Sweden to England which saved a lot of driving. We arrived in Norway with a station wagon full of gifts, and left with a car even fuller. It was nice to see our families again and the children got to know their grandparents better. Our parents of course spoiled the kids because they had not seen much of them. For example, there is a berry that grows on bogs in the mountains in Norway, called cloudberries in English. This berry is loved in Norway probably because it is rare and therefore only served at special parties. The cloudberries have a distinct taste, not really sweet like most berries, and our children did not like them. So they asked me: "Do we have to eat cloudberries, Dad?" I gave them no mercy: "Cloudberries are special in Norway and everybody loves them. You should learn to like them." The cloudberries have a position in Norway a little bit like goat cheese, which our kids often got on their sandwiches. When I went to public school in Norway I brought a lunchbox to school every day with two sandwiches containing goat cheese, and so did everybody else. Breakfast is normally included in hotel prices in Norway, and it contains all sorts of wonderful food like different fish, cheese, fruit, and vegetables. But you can recognize Norwegian customers because they always go over and take a few slices of goat cheese at the end of the meal.

We also went on a ski trip to the Alps in Switzerland during our stay in England. First we visited a Swiss scientist, Werner Kanzig, who used to work at GE. One of his students, Hans-Rudi Zeller, who as I mentioned, had been a postdoc with me for a couple of years was also present. It was interesting to me that when they spoke English together they used the first names Werner and Hans-Rudi, but when they switched into German it was Herr Professor Kanzig and Dr. Zeller. We stayed with the Kanzig family overnight. Next morning my car would not start. Kanzig being a practical fellow towed my car to the top of a tall hill in Zurich. Then I had a couple of miles to try to start the car in gear, but it still did not work. He then called a car patrol who fixed my car by changing the plugs.

After this we went to St. Morritz to ski. It was very different compared to in the US. You first take a gondola up into the mountains, and then further up the mountain skiers use T-bars as opposed to chairlifts. In the evening you would take the gondola down the mountain again. In the US you have lift operators helping you into the chairs or onto the T-bars. In Switzerland it was an automatic gate, and there was no one to help you. So the girls did not ski, but sat in the lodge playing a game called "pegged", while Inger, John, and I skied. However, my children tell me they did in fact ski as they remember several quite dramatic encounters with accelerating T-bars and falling off the lifts. When John rode with his sisters on the T-bars, he was so tall it hit him in the back of the knees, and he would fall while occasionally Anne or Guri were able to hang on without him. Guri in particular remembers the poma lift, where she was so light she floated in mid-air, trying to figure out how to jump off when she came to the end. My children's strong reaction to my recollection of our St. Moritz trip provides a clear example of what I alluded to at the very beginning of this book: namely that everybody's memories are somewhat different.

At the end of the day, we decided to ski down the mountain rather than take the gondola down, so Anne took her sisters with her in the gondola. As I recall when skiing down the mountain we came into a terrible fog, and had to ski from one stake to the next; they were spaced maybe 100 feet apart. So while Inger remained at one stake, John and I skied to the next one, then Inger followed. But one time when John and I reached the next pole, I did a little turn, and suddenly there was no ground under me. I fell into the side of the mountain and managed to put the ski-edge into the snow. When Inger arrived she said; "Where is dad?" and John calmly answered: "He fell off the mountain!" I in the meantime had stretched my poles upwards, and urged them to pull me up. It was so foggy that I could not see whether I was in grave danger or if it was only a small hill. We had found out before that the Swiss are very cavalier about safety compared to the USA. We decided to go back the next day to investigate what kind of danger I had been in, but of course we never did.

St. Moritz was a nice place and we managed to get sunburned because fog does not stop ultraviolet rays. We asked a Norwegian who worked there about the weather, and he said he had been working at the lodge for a month and a half and had not seen the sun yet. This is a deep secret

Guri and Trine posing at Stonehenge in 1970

of Switzerland, you really have to spend a long time there to see the sun. I had a chance to participate in Manfred Eigen's meeting in the Alps some time later, where we skied all day. The lectures started at 5 P.M. then dinner at 7 P.M. and then more lectures, but by this time only the speaker was awake. I skied with someone they called "Snow dog" off-piste and hurt my thumb when I fell. I had to go down to the village to see a medical doctor. The x-rays revealed that my tendons had been torn apart, and I needed an operation. Since I refused, he then set my hand in a cast holding onto a ski pole such that I could continue skiing the next day! But I never saw the sun this time either.

We went to Norway one last time before we returned to the USA, planning to sell the station wagon. In the US we had glued big flower-power stickers and peace signs all over the car. This was a popular thing to do in 1968, and the kids liked it. I went to the customs' department in Oslo to ask about the customs I had to pay for importing the station wagon.

and it was about $6000. By US standards, it should really have been sold for scrap by this time, but after it had been inspected by the customs' officials I was not permitted to discard it; that was the rule. I went back the next day, and got a different inspector. I asked him what people who are in my situation do. He said most people do not pay, but they unscrew what they can from the car and take it home with them. In a month or two the customs' service puts the car up for auction. You can then buy the car cheaply because of all the missing parts you have already taken. But if against all odds someone outbids you, you confront him and say "I happen to have most of the missing parts fitting the car you just bought. I will sell them to you cheaply." I told him I was going back to the USA within a week so that was not a solution. He took pity on me and inspected the car once more, said the underside was full of rust, and thankfully declared the car fit for scrap. I then sold it to a scrap dealer in Oslo for $500. That was not the end of the car however, because my sister Mette who lives in Oslo saw the car for at least two years after this. She could clearly recognize it because it was the only car in Oslo with big flowers all over it. In hindsight it really epitomized the obnoxious American tourist.

BACK IN THE USA

CHAPTER

14

Coming back to Schenectady the first thing we had to do was to buy another car, and we were fortunate that one of my friends, Bob Fleischer, lent us a Volkswagen Beetle for a week or so. We had rented the house out while we were away and that turned out to be a problem because the renter had a dog which we had not known about. Therefore Inger had quite a big job cleaning the house as mice had gotten into the dog food and even hid some of it in the furniture that we had stored in the basement.

Because of my stay in England I also got into problems with the IRS, and I was called into the office in Albany to explain my tax return. The office was in a basement in Albany and the receptionist was an older woman who had gnarled hands. My first thought was that she had been a tax cheater and had been tortured by the IRS, but she probably was an unfortunate woman with rheumatoid arthritis. Entering the office I was put into a booth, more or less like what you have in a men's toilet where the walls neither go up to the ceiling nor down to the floor. My examiner objected to the fact that I had deducted the private school tuition from my income. I lost that fight, even though I had to put the kids in private schools. The reason was that you cannot deduct the expense of going to Catholic schools in the USA. Next on the agenda was the expense of the tickets going to and from Europe; the Guggenheim Foundation had sent me a sheet stating that it was OK to deduct it, but the examiner refused to accept it. I had the receipts with me, and since I had also deducted the price of shipping the car, I kept shuffling that receipt underneath the others. Finally he said he would go and get his manager, and he left me alone in the room for a while.

In the room next door was a female examiner dealing with a carpenter who did not speak English very well. She had severe problems being understood by him and vice versa, and finally she said: "We had a similar problem last year, so I will let it go for now, but next year you must keep better records". Whereupon the guy in perfect English said thank you very much and that he would indeed do that.

Finally my examiner came back with his manager, who happened to be a woman. She examined the information I had gotten from the Guggenheim Foundation for a short while and then exclaimed: "They copied the law!" That is, of course, legal, so the tickets were deducted, no one cared about the car, and I could finally take my stuff and leave.

At the laboratory Charles Bean had changed his field to biophysics, and I decided to do the same, as I had studied a lot of biology while I was away. I first looked at an interesting problem where I extracted electrons with a high electrical field from a small metal tip. I sat in a dark room and watched the emission of electrons onto an electron multiplier and then onto a flourescent screen so they were relatively easy seen, but you had to adapt your eyes to the dark. This took time, so I decided to close one eye every time I had to go out of the room for some reason. When I entered the dark room again this became quite disturbing, because I could see nothing with the eye that had been open. I felt like I was blind in one eye.

Immunology

To work in biology the most difficult thing is to find a suitable problem to work on. Since I had always worked with thin metallic films in physics, I pretty soon decided to look at protein adsorption on metallic surfaces. To do this I reactivated an old ellipsometer, which is an instrument that can measure the thickness of films deposited on metallic surfaces. I did some preliminary experiments and found out that protein molecules basically adsorb in only a single layer on any metallic surface, after which I decided to try immunology. I had been introduced to immunology in an undergraduate course I took in England. I got hold of some bovine serum albumen (BSA) and serum from a rabbit containing antibodies to BSA. A BSA protein molecule in biology serves the same function as the hydrogen atom in physics, i.e. it is very basic. I first adsorbed bovine albumin from a saline solution onto a gold film that was evaporated onto a microscope glass slide, while measuring the thickness. It grew from zero to about 30 angstrom and stopped. When I exposed this film to rabbit serum containing antibodies, the film would increase another 50 angstrom or so, which meant that I now had a double layer of protein on the surface. The basic rules of protein adsorption are that protein will adsorb on most surfaces but not on another protein unless the protein is the correct antibody. This worked very well, was very reproducible, and could in principle be used to test for illnesses. It did not take a long time to recognize that the expensive ellipsometer was unnecessary to detect single and double layers of proteins; we could simply use interference colors. Bismuth oxide worked very well, but was not stable in a saline solution and the serum is a little salty. Tantalum oxide was my next try; it worked well and was stable. So I applied for a patent on my system to detect antibodies. But a little later I learned that a scientist named Alexsandre Rothen at Rockefeller University already had done similar experiments. The patent people went ahead anyway because Rothen had only showed that it worked, but not said that it could be used for clinical purposes. So against my wishes they applied for a patent and I got it. I later found out that indium that had been evaporated slowly onto a glass slide made an even better substrate to detect thin films. The reason is that indium will agglutinate into small particles and scatter light very efficiently. By absorbing protein on the particles the light scattering changes dramatically and the protein layer can be easily detected with the naked eye. By subjecting this layer to the correct antibody, the color changes again.

I wanted to show that I could detect antibodies to a real disease and looked around in the Capital District for a coworker. I first went to a laboratory in Slingerland that specialized in rabies. To work there I had to be vaccinated against rabies as a precaution, and I decided against that as rabies is a deadly disease. Eventually I found an excellent teacher, Dr. Laffin, at the Albany Medical Center, who was interested in collaborating with me. One interesting problem he worked on was detecting hepatitis and after conferring with Dr. Laffin, we decided to try that. The GE Laboratory was maybe a half hour drive from the medical center so that was no problem, but parking was a pain. Nonetheless I was willing to put up with that. We worked together for maybe a short year and published an article in *Proceedings of the National Academy of Science* (PNAS). The article dealt with detecting antibodies to hepatitis. But then this program was interrupted, because on October 23, 1973, I received the Nobel Prize.

CHAPTER 15

RECEIVING THE NOBEL PRIZE

The GE Research Laboratory had started a program similar to a Ph.D. program. With a bachelor's degree you could be hired at the Lab for two years and work for a scientist while simultaneously studying at a nearby university. I had helped to hire a young Swede into the program, John Valin, in 1971. I had met him in Sweden when his wife was pregnant, and he hoped that it would be a boy, because they had two girls from before. So I bet him a bottle of cognac it would be. I knew the odds were slightly on my side, as there are approximately 2% more boys born than girls. It was a boy, and he showed up at my house one day with the bottle. We became good friends and played a lot of tennis together. After two years he went back to Sweden, but he remained in contact with my new manager Dr. Milan Fiske. I had no idea, but I had been suggested for the Nobel Prize by people at the laboratory. Actually they needed help, because only previous Nobel Prize winners, professors at Scandinavian universities, or universities in the world selected by the Nobel Prize committee can suggest people for the Nobel Prize. So I have no idea who submitted the application. There had been a tunneling conference in Riso, Denmark, in June 1967, arranged by Professor Eli Burnstein. That fall I was a little nervous because it was rumored that I had been suggested for the Nobel Prize. Since I did not get it at that time, I forgot about it. In my mind I thought I would have a chance if superconducting tunneling became practical. The reason it did not is, of course, that it requires low temperatures. Anyway this time Milan used John Valin as his spy in Sweden since he was at Chalmers Technological University where two of the professors on the Physics Nobel Prize Committee worked. On October 22, 1973, Milan called me into his office and said he had heard there was a chance that I would receive the Nobel Prize in Physics the next day. The reason he did this was that my mother was again visiting from Norway and he knew she was scheduled to leave the next day. He thought it would have been awful for her to learn that her son had received the Nobel Prize after she got back home in Norway. But he emphasized that he was not certain. Milan was a very considerate person.

Charlie Bean walked me home that day; I live quite close to the GE lab. He of course knew that I had been suggested. He also knew that they had asked NTH, my alma mater, to support me, but the only letter they got from Norway was that somebody with my name had graduated from NTH in 1952 in mechanical engineering with bad grades. Charlie and I talked about the possibility of me winning, and Charlie put it at 50%. I said if I won I cannot be fired, not expected to be fired anyway. When I got home I had to tell my mother about the rumor and ask her to delay her return to Norway. Of course I also told Inger about the news. The kids knew something was going on, but I thought it best not to inform them. That evening we went to a local restaurant, and ate roasted duck. I remember it well because Inger and I seldom eat out.

At noon in Stockholm, on November 23, 1973, the head of the Nobel Physics Prize committee appeared on the doorstep of the Royal Swedish Academy of Science and announced who had won the Nobel Prize in

Physics. John Valin was waiting and immediately called Milan who then called me at home one minute after 6 A.M. He said only: "Ivar, congratulations, you made it!" It was music to my ears, but since I had been properly warned, I did not get the same kick as I got receiving the Buckley Prize. I got a big hug from Inger who was standing beside me. When I put down the phone, it immediately rang again, and now a person with a Swedish accent congratulated me, and hopefully I acted surprised. As soon as this conversation ended, the phone rang again and this time it was a newspaper. In the meantime my mother had come into our bedroom and we all hugged each other. I then called out to the kids; Anne and Guri were excited, but Trine was probably too young to understand the significance. John had left home three years ago to go to Cornell, where he stayed for two years. Then he dropped out, bought a big motorcycle, and travelled around the US. He was taken by the movie "Easy Rider." At the time he was someplace in the state of Washington working for the forest service.

Ivar and coworkers in GE hallway, Oct 23 1973

We ignored the phone after that, ate breakfast, and then the kids went to school as usual. Anne, I remember, was having her SAT test that day. About 8:00 AM a big GE limousine came into my driveway. I usually walked to work, so I went out and dismissed the driver. Ten minutes later he came back, and this time he begged me to get in the car, saying it was a request by the director, Art Beuche. So this time I could not refuse. It took maybe two minutes to drive to work, and when we got there they had a red carpet on the ground outside the laboratory that I could walk on to get inside. Standing there were all my colleagues and my manager Milan Fiske. As for me it was complete chaos to try to keep my wits about me. A press conference had already been arranged in the auditorium where I had first spoken about my tunneling experiments 13 years ago. Then someone had informed Mr. Reginald Jones, the chairman of GE, who called me. This was October 23, 1973, which happened to be William Coolidge's 100th birthday. Coolidge was a very famous GE scientist, who had made many inventions. One was the ductile tungsten which is used in all incandescent light bulbs; another was the practical x-ray tube.

Press conference at GE Lab Oct 23 1973

Party at GE Laboratory celebrating Ivar's Nobel Prize, Oct 23 1973

Ivar and his three daughters, Oct 23 1973

He really had deserved the Nobel Prize! Because of his birthday GE had already planned a big celebration and had stocked up on lots of champagne. GE decided to postpone Coolidge's birthday celebrations by one day and celebrated me instead.

It was a nice party, and the whole laboratory assembled in the lobby for champagne, snacks, and to shake my hand. My whole family was there except John. He contacted us a day or two later and decided to come home. I was particularly happy that my mother was there and could experience this. GE had found a Norwegian flag and put a big sign up which said simply: "Congratulations Ivar" and my mother appreciated the informality.

Both Inger and I were in doubt about how we should behave as we had little time to talk together. But we did manage to arrange a party for

Inger and Ivar with Anne, Guri and Trine, Oct 23 1973

all our neighbors that evening. GE was generous and supplied us with more champagne and exotic food because we had no chance to go to the store. By about 10 P.M. we finally were alone, beat and ready for bed. But not my mother! She wanted to see the 11 o'clock news because I had been interviewed on TV during the day. I tried to persuade her and said: "Mother I am here, you can talk to me in person", but to no avail. So we all saw the 11 o'clock TV news and were in bed by midnight.

The next day I was late at work because I was buried in telegrams and various best wishes from the day before. I particular enjoyed a telegram from my old poker friend Kristen in Norway because the address said simply "Ivar, USA" and I received it! I wish everybody could have such a day. It reminded me of Warhol's quote that "in the future, everyone will be world-famous for 15 minutes". In the US as

My mother with John who came home a few days after Oct 23

well as in Norway, the Nobel Prize does not give you everlasting fame, especially when compared to the esteem bestowed on Nobel Prize winners in China, Korea or Japan. Leo Esaki, who is Japanese, and Brian Josephson, who is Welsh, shared the Prize with me. Brian got half, and Leo and I shared the other half. Leo later became President of the University of Tsukuba for example, while I was only offered a regular professorship at NTH in Norway at the fixed professor salary. That evening I had the opportunity to meet Dr Coolidge, and to apologize to him for having stolen his party. But he smiled and forgave me. He had just received an artificial diamond from GE that he admired. Remember he was 100 years and 1 day old!

My former brother-in-law in Norway, Roald Aass, won a gold medal in speed skating in the 1960 Olympics in Squaw Valley and he, in contrast to me, will remain forever famous. Maybe 10 years after I received the Nobel Prize and about 25 years after Roald had won the gold medal, I drove with Roald to see a soccer match in Oslo. At the stadium there were no parking places, and Roald rolled down the window of his car to ask a policeman for parking. Before he could say anything the policeman said: "But is it really you, Roald?" Roald answered: "Yes, it is me, but what shall I do with my car?" "Just leave it right here, I will look after it for you!" the policeman answered.

Now started a busy time for me because I recevied lots of invitations to give talks, but I also had to prepare to go to Stockholm to receive the Nobel Prize. The Nobel Prize is always given out on December tenth, the day Alfred Nobel died. At the time Professor John Bardeen was consulting with GE and I asked him how I should behave when receiving the prize. John said he would write me a letter. A few days later I received a letter from John that started:

"Dear Ivar,
The *first* time I got the Nobel Prize; I did not bring my family and regretted it."

Professor John Bardeen and Ivar in November 1973

At the time the Nobel Foundation only paid for travel and hotel accommodations for the recipient and not for the family. John Bardeen convinced me to take Inger and all four kids. It turned out that GE ended up paying for all of us in the end anyway. I also invited Inger's parents and my mother from Norway. My father had died a few years earlier. Inger's father decided to stay home; he unfortunately had polio as a young man and had difficulty walking.

Talks in Norway

We first travelled to Norway where I gave a talk about immunology and the indium slides rather than the superconducting tunneling work for which I had won the Nobel Prize. I had a large audience, and I asked if anyone was present that did not speak Norwegian. Only one person raised his hand. I apologized to him and gave my talk in Norwegian.

Then we went to Trondheim where NTH is located. Arthur Sarsten, who used to work for Alco in Schenectady and was a good friend, was on the invitation committee. He was now a professor at NTH. We had decided to fly, but he insisted that we had to take the train. When we arrived at the station in Trondheim we were met by a marching band who welcomed us. We stayed with Arthur and his wife, Solveig, when we were there, rather than going to a hotel.

I gave a talk about immunology again the next day, and at night NTH held a party for me. I met some of my professors from 25 years ago. I was particularly nervous to meet Professor Hole who had been my physics professor. At the time I thought I might have flunked physics and remember that I went to ask Professor Hole how I had done. We did not write our names on the answer sheets, only the seat where we sat. So he said: "Where did you sit, Giaever?" I answered: "Seat number 2 row 4". Looking down at his notes he said: "Your answers are among the worst that have ever been handed in at NTH!" In Norway the professor teaching a course does not set the grade, but another professor at another university does. Therefore I did not flunk, but managed to get the lowest passing grade of 4.0. Anyway, when I met him I realized that he was more anxious than I, and we had a cordial chat.

Stockholm

From here we visited both Inger's family and my family before we headed for Stockholm by train. The head of the Nobel Prize foundation met us at the train station and drove us in a brand new Volvo to the Grand Hotel in Stockholm that is used by the Nobel Foundation every year for this purpose. Our mothers also stayed there. I was assigned a Volvo and an adjutant who I believe was a Swedish graduate student. I had rented tails in the USA before I left for Sweden, and the guy asked me what instrument I played. Only musicians use tails in the US. I told him I did not play any instrument, but that I was about to meet the King of Sweden. He thought I was joking. Now I used this opportunity to ask my adjutant if my rented tails were OK. He said they were passable, but recommended that I rent another one in Sweden if I wanted to look sharp. And I since I definitely wanted to look sharp, I did.

On the morning of December 10, we went to the Concert Hall to actually receive the Prize, consisting of a diploma and a medal; the money was to be received later through a bank. This year the ceremony was particularly interesting for the Swedes because it was the first time the new king, Carl XVI Gustave, who was just 27 years old, was giving out the Nobel Prizes. We were told what to do and there are rules: the order follows the order of Nobel's testament. The oldest physicist gets the prize first, followed by the next oldest, and so on. Then it is chemistry's turn, followed by physiology and medicine, then literature, and finally economics. The Economics Prize is not technically a Nobel Prize, but rather, a prize given by the Swedish banks administered by the Nobel Foundation. But if you ever meet one of these people you should not tell him or her that!

After the audience was settled, there was fanfare announcing the arrival of the Swedish king with members of the royal family. Then the Nobel Prize winners came in a row with the oldest physicist first. Each

Ivar receiving the Nobel Medal from the Swedish King

winner was accompanied by a member of Swedish Academy. When we were settled on the stage, everybody stood up and sang the Swedish Royal Anthem. After that a piece of music was played by a full orchestra and finally the award ceremony began. Leo, as the oldest, was the first to receive the Prize and after that it was my turn. Professor Lundquist, who headed the Physics Nobel Committee that year, first told the audience in a few minutes what I had done to deserve the prize. The king then handed me the diploma and medal and we exchanged a few words. I shook his hand, turned to bow to the audience who were clapping, and went back to my seat. In Sweden they all wanted to know what I said to the king. It is supposed to be a secret but this is what I said to him in Swedish: "It is nice to meet you on this occasion." First he was a little puzzled, but then he said that he knew he could speak Norwegian with me; Norwegian and Swedish are quite similar languages.

Later that day we went to the City Hall for the Nobel banquet. I had been warned by John Valin that I would probably have to give a talk at that dinner. I had carefully prepared for it long beforehand. But then we were told by the toastmaster for the evening that he had studied some Japanese phrases and wanted to introduce Leo Esaki. Leo had not prepared, so I gave my speech to him. I did not like his delivery, but being there I recognized that my speech was too serious and dull anyway.

There are strict rules for the seating at the table as well. Each woman is assigned an escort. Mrs. Esaki, who was married to the oldest physicist, sat with the King. Inger, married to the next oldest, had Prince Bertil as a cavalier, while I escorted Mrs. Wilkinson at the table, the wife of one of the Chemistry Nobel Prize winners. Brian Josephson was seated across from me, and he was one of the favorite subjects of the photographers. There was only a brief pause when pictures were allowed to be taken. Brian had a small box camera where you attached a flashbulb (cube) on the top. I took a picture of him and he tried to take a picture of me, but then the flashbulb popped out of the camera and fell under the table. That did not faze Brian; he simply disappeared under the table and was absent for a long time, before he triumphantly came up with the bulb in his hand.

After dinner there was dancing, and being polite I danced the first dance with Mrs. Wilkinson. I had been told that the first dance belonged to your tablemate. Inger did not like that because Prince Bertil did not dance. Later there were lots of pictures in the newspapers where Mrs. Wikinson was identified as Inger because she danced with me.

All in all the short week in Stockholm was very enjoyable and we went to several dinners. One was held at the castle where Inger got the Swedish Prime Minister Olof Palme at the table, and I qualified for Princess Christina. The rule at the Castle is that there are no speeches so it was a relaxing evening.

On December 13 we were woken up at The Grand Hotel by 6 or 7 young girls with crowns of burning candles in their hair. They were all dressed in white and looked like the angels I had on my childhood bed. They were singing a beautiful song in Swedish. This was on St. Lucia day, one of the unique traditions in Sweden. At that time people were not so safety conscious; I have been back in Sweden since and now the candles are all electric.

All the Nobel winners were invited to participate in a TV program, where we just talked in general. I enjoyed that because Nicolaas Tinbergen, a Physiology and Medicine winner, was smoking while he complained about pollution, and I got some smiles for pointing that out. The producer of the show was very proud of the fact that they always showed it with no retakes or edits. There was only one exception when Lars Onsager, a Norwegian-born Nobel Prize winner was on the show. Turned out Onsager fell asleep, then he suddenly woke up, stretched out his arm and said: "A cup of tea, thank you." They edited that one out.

We also had to give a talk at Stockholm University about the Nobel Prize work. I remember I decided to include the Josephson Effect in my talk. I had actually published a paper before the Josephson Effect was known which showed the DC Josephson current. But of course I did not recognize it at the time. It is not enough to measure something, you also have to know what you are measuring! I have been asked many times if I am sorry that I missed out on the Josephson Effect, but how can I be sorry when I got the Nobel Prize? It is required to turn in the text of your Nobel Prize lecture to the Nobel foundation, which I did. I have been told that Murray Gellman is the only person who did not supply a written copy of his speech, but since he is really famous, the Swedes must have forgiven him.

Before we left Stockholm, Inger and I bought some Swedish crystal vases for the people we suspected had supported me for the Prize and had them shipped directly to the US. We said a nice farewell to the other Nobel laureates and in particular to Leo and Brian. Brian introduced me to his parents; his mother was a teacher for gifted children! Brian wanted to know how I would get the Nobel Medal into the US and I said I would carry it in my briefcase. He was worried that he would have to pay a customs' fee. I said I didn't think there would be any, but I did not know. He said he did not want to pay the customs' tax, so he would put it into his shoe. I said you got half the Nobel Prize money, so you can afford to pay if you have to.

Copenhagen

From Stockholm the whole family went to Copenhagen as I had been asked to give a talk at the Bohr Institute. I met Aage Bohr there, a Nobel laureate and the son of famous Niels Bohr. Aage was a good host and invited us home for a family dinner. We also went sightseeing in Copenhagen for a couple of days, visiting the Little Mermaid statue in the harbor and then onward for some shopping in Den Permanente, a famous exhibition in Copenhagen. Inger and the children went home after the visit to Copenhagen while I went back to Sweden to give a talk at Lund University.

A couple of days later, I took Scandinavian Airline System (SAS) back to Kennedy airport. On the airplane I was recognized by Kirstin

Salk who had stayed in The Grand Hotel in Stockholm as well. We got talking and I told her that I had won the Coolidge prize, the highest accolade GE lab can give an employee. The prize allows you to go to an institution for a year to learn something new, and I wanted to study more biology. It turned out that Kirstin Salk was Jonas Salk's sister-in-law and she highly recommended the Salk Institute in La Jolla. I had never heard about it, but of course I knew who Jonas Salk was, and the suggestion intrigued me. She offered to put in a good word for me, and I spent part of the time there on my sabbatical Coolidge fellowship.

When I got back to Kennedy airport the customs' official asked me if I had acquired anything abroad. I opened the briefcase and showed him the medal. He immediately called out to the other nearby officials: "Hey Fred, Bob and George. Come and take a look at a Nobel Prize Medal!" No talk about paying customs, but in England it might have been different.

Back to Niskayuna

It was sort of anticlimactic to get back to Schenectady, but it was right before Christmas and we had to get ready with presents for the children and also get a Christmas tree. We still celebrated Christmas the Norwegian way. On Christmas Eve we celebrate by singing while dancing hand in hand around the Christmas tree and giving out presents. The youngest person gets the first present, then the next youngest, and so on. This year was special for many reasons. First we had gotten into some money. I put the money into the bank in the children's names to be used for their college education. It was not all that much money, 25% of the Nobel Prize was roughly $30,000 tax free. At the time the value of the Nobel Prize was almost a minimum. If I had gotten the prize in 2014 I would have received about $400,000 before taxes. The USA is now the only country that taxes the Nobel Prize. All in all I was not lucky with the timing of my Nobel Prize. My original plan to pay for my children's college education was to pay for John, and then he would help pay for Anne since he was 4 years

Inger and Ivar 1974 participating in a film about tunneling

older than her, and Anne would pay for Guri, and so on. Since John had dropped out of college that plan was no longer viable. John had actually finished a bachelor's degree in liberal arts at the University of Oregon. This Christmas he wished for a Texas Instruments calculator, probably because my Nobel Prize inspired him to start at the University of Washington to finish an engineering degree. He had studied engineering the 2 years he was at Cornell.

That winter I did not do much science; my suitcase was always packed. I received invitations to give talks everywhere, and I did not want to appear arrogant or snooty, so I accepted most of them. At first I enjoyed giving the talks, but it became routine pretty soon, and I wanted to go back to my work in biology.

When I received the Nobel Prize, I was elected to the National Academy of Science, and one privilege as a member was that you could publish papers in PNAS without an extensive review. The only academy member who has ever been refused for publication to date is Linus Pauling; he wanted to publish a paper that claimed that vitamin

C prevented and cured cancer, but an editor stopped that paper. When I met Pauling, he was rather proud that he was the only one who had been refused.

Thoughts about the Nobel Prize

As a Nobel laureate you get lots of invitations to travel and it is tempting to accept more invitations than you should. Actually it was not as common to get offered money to give talks in 1973 as it is now in 2016. For example, in 1978 GE celebrated its 100th anniversary which came as a surprise to the employees, because 8 years before, in 1970, GE had celebrated its 75th anniversary. The reason for that was that in 1895 several small electrical companies including Edison Electric got together and formed General Electric. But in 1978 General Electric decided it had been formed by Edison himself in 1878, obviously for advertising purposes, and thus changed its history.

At the General Electric Research Laboratory we celebrated this occasion by inviting several Nobel Prize winners to a general scientific conference. I happen to know that Dr. Francis Crick was offered $1000 and free economy class airfare to come from California to GE in New York. He refused and said he would come for $3000, which we ended up accepting. These days it is common for Nobel Prize winners to get offers of $5,000–$10,000 to attend a meeting and give a talk, plus business class travel worth a comparable amount. The focus of Crick's talk was about introns and exons which had just been discovered but not by him, and afterward he was worried because it had not been published yet. Scientists at the laboratory displayed their work informally, and the visitors went around and looked at the exhibits. I was working with the indium slides at that time, and remember Crick and Watson coming together into my lab and looking it over. I have no memory about how they liked it, so they were probably just polite. Another Nobel laureate visitor was Professor Melvin Kelvin who talked about getting gasoline from rubber trees. The reason he talked about this was that the price of oil during the OPEC embargo rose from $3 a barrel in 1973 to over $12 a barrel in 1978, and continued to rise. He got objections from a chemist in the audience who pointed out that long molecules — rubber — were more valuable than short molecules — gasoline. And now that the price is over $100 per barrel no one talks about that idea. Good scientists often come up with crazy ideas at the end of their lives. Linus Pauling is a good example. He believed that vitamin C both cured and prevented cancer. The first time I was in China I was asked on two occasions what people in the US thought about that. That showed how famous Pauling was. I met Pauling a couple of times and told him that I now worked in immunology. He told me he had contributed to immunology. I remember that I told him that I knew that and that he had been wrong. He was not too happy about this answer, but of course he knew that too. He had postulated that antibodies were modified by the antigen, which was not an unreasonable suggestion at the time.

I am a climate skeptic or denialist, is that my own crazy idea? At the present time climate change is on everybody's lips, and it is supposed to be caused by global warming. According to published record, the highest global average temperature measured was in 1998. We are now in 2015 as I write this, 17 years later, and even with lots of more CO_2 in the atmosphere, it is not as warm as it was in 1998. The concentration of CO_2 was 295 parts per million (ppm) in 1898 and increased to 367 ppm in 1998, i.e. it increased by 72 ppm while the temperature rose by 0.8 degrees Celsius. From 1998 to now, the CO_2 has increased to 403 ppm, i.e. another 36 ppm or half the total of the previous 100 years, but the temperature is constant. Since the global warming people believe that CO_2 concentration caused the rise, the measurements do not support this. Therefore everybody now talks about climate change, and keeps quiet about the lack of global warming. If money were not desperately needed for example for the refugees from Syria, I would not mind people building solar cells or windmills, but it is hard to see so much waste. Since the global warming people could advocate for nuclear energy, which emits no CO_2, it is hard to understand their logic. I believe that maybe 50 years from now people will ask, what in the world was wrong with us, not wanting to use cheap energy, which is the real reason for our growing prosperity?

CHAPTER 16

MY FIRST TRIP TO CHINA

On a different adventure entirely, I was invited via mail by Professor Charles Schlicter, a professor at the University of Illinois in Urbana, to go to China for one month together with some of the most famous solid state scientists in the USA. I had travelled a lot that year, and since Inger could not come along, I said no to the trip. But then I met Schlicter at an American Physical Society meeting and he asked me in person. Since he was very persuasive, I changed my mind and said yes. I have never regretted that decision.

We were a group of 12 people altogether, most of us physicists:

JOHN BARDEEN, Professor (emeritus) of Physics and Electrical Engineering and in the Center for Advanced Study, University of Illinois, Champaign-Urbana, Illinois
NICOLAAS BLOEMBERGEN, Rumford Professor of Physics and Gordon McKay Professor of Applied Physics, Harvard University, Cambridge, Massachusetts
LEROY L. CHANG, Member of Research Staff, Thomas J. Watson Research Center, IBM Corporation, Yorktown Heights, New York
SAMUEL C. CHU, Professor of History and Director of the East Asian Program, Ohio State University, Columbus, Ohio
ANNE FITZGERALD, Professional Associate, Committee on Scholarly Communication with the People's Republic of China, 2101 Constitution Avenue, N.W., Washington, D.C.
THEODORE H. GEBALLE, Chairman, Department of Applied Physics, Stanford University, Stanford, California
IVAR GIAEVER, Biophysicist and Coolidge Fellow, Physical Science Branch, Physical Science and Engineering, General Electric Research and Development Center, Schenectady, New York
JOHN J. GILMAN, Director, Materials Research Center, Allied Chemical Corporation, Park Avenue and Columbia Road, Morristown, New Jersey
W. CONYERS HERRING, Member of Technical Staff, Bell Telephone Laboratories, Murray Hill, New Jersey
J. ROBERT SCHRIEFFER, Mary Amanda Wood Professor of Physics, University of Pennsylvania, Philadelphia, Pennsylvania
ROBERT H. SILSBEE, Professor of Physics and Director, Laboratory of Atomic and Solid State Physics, 517 Clark Hall, Cornell University, Ithaca, New York
CHARLES P. SLICHTER, Chairman of the Delegation, Professor of Physics and in the Center for Advanced Study, University of Illinois, Champaign-Urbana, Illinois

Our group posing on Chinese Wall

The purpose of the trip was to evaluate the Chinese effort in solid state physics and get a feel for how well they were doing. It was probably also to establish normal relations with China, which President Nixon had started with his visit in February 1972. The first visitors sent to China were scientists and ping-pong players because they are not very political and are therefore very safe visitors.

We had a meeting in Washington before we left, and a biology group concerned with a disease called schistosomiasis clued us in. At the time I had never heard about schistosomiasis, but I would later study the immunology of the disease. The disease comes from a parasite that lives in a fresh-water snail. The parasite can enter through people's skin if they are wading in the water, and eventually ends up in the portal vein as a worm. Then it becomes an egg-laying machine and causes people to be sick, while the parasite's eggs reenter the water through feces or urine. The researchers were very disappointed with their trip, because Mao had declared that the snails no longer resided in China, and had closed all the schistosomiasis facilities.

We flew first to Tokyo after a brief stopover in Alaska, and spent a night in Japan. The next day we arrived in Beijing where we spent 10 days visiting various institutes and universities, talking mainly about solid state physics. I found the scientists very knowledgeable. They had developed mostly the same tools and instruments that we had in the US, and had done it without any help from outside China. This was at the end of the Cultural Revolution, so most of the universities we visited had been closed for a couple of years, but the Chinese did not like to acknowledge this fact.

The first night we stayed in Beijing I was woken up in the middle of the night, maybe around 2 A.M. by noise from the street and went outside to investigate. There were banners all along the avenue, and people dressed in festive clothes with many children marching. I tried to take a picture, but a hotel employee prevented me from doing so. I asked him what was going on and he said: "It is a spontaneous uprising to wish Prince Sihanouk farewell. He is leaving for the Khmer Republic." The next morning at breakfast I asked my friends if they had attended the show last night, but they had not heard a thing, and looked like they did not believe me. I said proudly that they will see the evidence out on the street. When we got out, there was nothing left on the street, all the banners had been taken down and all the people were back in the drab Mao uniforms.

We were joined by a Chinese interpreter in Beijing. She knew written English very well, but not spoken English. I enjoyed trying to talk with her, and I asked her whether she ever had acupuncture. She said no, but her brother had been treated twice. I asked what was wrong with him. She said: "He did not like to work in the field!" I asked if it helped, and she said: "Absolutely!" I had a friendly bet with Ted Geballe, a professor from Stanford, that acupuncture does not work. Thus I bought a set of acupuncture needles with the charts both in Beijing and Shanghai and discovered that the places you insert the needles were different in the two cities. I have since learned that any acupuncture practitioner has his or her own points where the needles are inserted. I once asked most of the Chinese students at RPI if they believed in acupuncture and they all did. It is really a cultural bias, since the Chinese have been told from childhood that acupuncture is a good thing, and they never question it. There is absolutely no scientific evidence that acupuncture works, so it is hard for me to understand that at this time many people in the USA also believe in the procedure. Maybe medical insurance is the culprit. If you have backache you might have an x-ray or an MRI scan done which is expensive. If the backache persists the medical insurance company will suggest acupuncture, which does not heal you either, but at a medical cost that is just a fraction of an MRI scan. It is similar to water dowsing which many people in many different countries swear by, even in Norway, but it absolutely does not work either.

We had many discussions with older scientists who had been educated in the USA and spoke English, but the discussions always went through an interpreter. The discussions were very formal and we were always served green tea. The venues were also very old-fashioned rooms with big chairs, and there were always a few spittoons on the floor. The Chinese spat a lot at that time, and when you heard bicyclists behind you cleaning their throats, you were always afraid of being hit by the spit. We went to see the Forbidden City which was full of visitors, but as soon as we approached the Chinese withdrew; we could never get any contact. They must have been told not to have any contact with foreigners. And if I went out and stood in the street, maybe a hundred or so Chinese people would gather and just stand there watching me maybe a few hundred feet away. They were not used to seeing people from the West. At the Forbidden City we were shown lots of beautiful antiques, among other things a famous sculpture of a horse. The interpreter emphasized that everything was real. Ted Geballe remarked that that was strange because he had recently seen the very same horse in San Francisco, as it was on a tour in the US. Some confusion ensued, but in the end we were told that everything was real except the horse!

I had arranged before we arrived in Beijing to have a "three cushion billiard" match with John Bardeen and Nico Blombergen. We all looked forward to the match, had talked it up, and when we went down to the basement of the Beijing Hotel where the billiard tables stood, we had many spectators. Bardeen took the first shot; the ball hit the first cushion and stopped dead. We could never finish the match, in fact we could not even start, because all the cushions were all dead.

We traveled mainly by trains inside China and I remember going to Nanking. I like to play games, and on the train I learned to play Chinese chess, not checkers. The game is similar, but different from regular chess. They have some pieces called "cannons," which are similar to a rook in chess, except that they can actually jump over other pieces. We

Bloembergen and Bardeen at the billiard table in Hotel Beijing

had two Chinese people following us on the train trip — a man and a woman. I ended up playing a chess game with her. She was supposed to be a novice like myself, but I ended up losing badly. I was later told by our interpreter that the man was very good, and he guided her during the game. So communists cheat as well. We were told that men and women were treated as equals in China because that is the proper communistic rule. But we also had the opportunity to travel by air a few times, and I noticed that the passengers were mainly men. They used the kind of small airplanes where the seats are under the wings. The only women I saw were the stewardesses who came into the cabin to check that the wheels were properly down before we landed.

We ended our tour in Shanghai, and got to our hotel the day before October 1, which is Independence Day in China. At the time I went running every morning and had continued that in China. I enjoyed seeing the Chinese doing their slow 'balancing exercises' in the parks. In Beijing I

Curious Chinese children

had run with Conyers Herring a few times for company, and when we got back I tried to discuss with him what we had seen on the trip. But he had been concentrating so much on the running that he had hardly noticed anything. This time I ran alone and maybe I should have concentrated more on the road, because in Shanghai I could not find my way back to the hotel. Worse, I did not know the name of the hotel as only Chinese characters were used. Fortunately, they had policemen standing on the main corners directing the bicycle traffic as there were almost no cars. So I approached one of the policemen and tried to indicate that I was lost. It was a little like playing the game of charades. Very soon I had gathered a few hundred people that took part in a discussion that to me was a complete mystery. Since no one was actually helping me, I walked away and tried the next corner, and had the same experience. This time when I walked away I passed by the Polish Embassy. I was tempted to walk in because at least I could find somebody there who spoke English. But this was during the height of the Cold War, so I decided to continue on my way. During the next game of charades two very young girls took me by the hand and guided me to a hotel where they spoke English. I was at first not understood, then I said "American delegation" and the person said "Your name: American delegation?" I said "No, no telephone." So he handed me a telephone, but of course I could not use it. But then he got smart and called some place where they had our passports. The US passport service has a rule: "Never give away your passports," but in China we did not have a choice. They of course knew the hotel I was staying at, and they sent a car to get me to the correct hotel. When I got there the delegation was outside ready to partake in the Independence Days of celebrations, but they said they would wait for me. I rushed up to my room, got dressed, and grabbed my camera to put in a roll of film. In my hurry, I made a mistake loading the film and as a result I have no pictures from that day.

In China everybody dressed in a drab Mao uniform and cap, so all the Chinese people looked alike. But somehow you could tell that the women in Shanghai were different from the men, though in Beijing they looked the same. Or maybe it was because I had been in China for a whole month by now and missed female company. I am not a very tall person, but in China I was taller than most people we met on the street. But we had the opportunity to watch the opening of the National Games in Shanghai. This was a sporting event. The Chinese that came marching into the stadium were very tall, so you recognized that they had been selected from the population of 1 billion people. Another thing I noted was that the different provinces marched in "alphabetical" order; the province with the fewest strokes in their Chinese name came first and so on. I have been back to China many times since, and it is impossible to understand how the country could have developed so much over such a short period of time. In 1975 Beijing was a village with several millions of bicycles and 'night soil' was carried in the streets. Today it is a city of buildings 20–30 stories high, and it is difficult to cross the streets because of the car traffic. How they managed this economic miracle is difficult to understand, but the fact is that they did. To me it is unbelievable because the speed in Norway is much slower; it took maybe 10 years just to build an opera house in Oslo.

CHAPTER 17
WORKING IN BIOLOGY

Working with schistosomiasis

Strangely enough, shortly after I learned about the disease schistosomiasis, when I attended the session to prepare us for the Chinese visit, I would be working with it. Someone working at the Winches Farm in St. Albans, England, contacted me after I had published a test for hepatitis with the indium slides and wondered if it could detect antibodies for schistosomiasis. I said I was sure it would work. Then I had to prove it, because they asked me to participate in an international test arranged by the Rockefeller Foundation. This was very late in the day to enter the contest, but I managed to apply to participate before the deadline and was accepted. The Winches Farm promised to give me the proper antigens.

It turned out I had to do the test at the Hospital for Tropical Diseases in London, because I could not bring the specimens to the GE Laboratory. The specimens consisted of maybe 100 (I can no longer remember) vials with 5 milliliters of blood collected from two or three different places in Africa. So I went to London and checked in at a small bed and breakfast room across from the hospital. First I had to go to the Winches Farm in St. Albans to pick up the antigens. At the laboratory they were very forthcoming and interested in my immunology test. They also showed me around the facilities, which were not very modern, and that is typical English. When they went for my antigens, they said sit here and wait for a while, and they gave me a cup of coffee. I asked the guy sitting with me if he should not have coffee as well, but he said: "We are not allowed to eat or drink in the laboratory!"

I got my antigens and the next morning I went to the hospital to do the tests. Somebody at the hospital showed me a room where I could work and the refrigerator where the specimens were stored. I had plenty of microscope glass slides covered with indium with me, and to fit the slides into the vials I had to cut them into 4 equal squares with a glass cutter. Then I put two small drops of antigens on each slide and waited for half an hour. Then I mouth pipetted about 2 milliliters of blood from some of the specimens and transferred the serum into new vials. Next I rinsed off the antigen from the slides, and incubated the slides for 2 hours with a twofold dilution of the serum. Then the indium slides were rinsed, blown dry, and examined. If the spots could be clearly seen the specimen was positive, if they were not visible the specimen was negative. I was a little reluctant to use my mouth to transfer the blood, but had no choice. The hospital was surprisingly rather dirty and old fashioned. This was before AIDs had become a big thing, and had I known about AIDS I would never have transferred blood from Africa by mouth. While I did the tests I was observed by many scientists in the laboratory; they were impressed by my results, but they thought the test was too awkward, in particular the cutting of the microscope glass-slides with a glasscutter! Somehow they did not recognize that I just did what was convenient at the time. If the test became commercial, everything would have been properly prepared in a factory. Anyway they were nice people and I took them all out to a restaurant for Indian curry before I left.

The Rockefeller Foundation invited us to a place in Massachusetts where we would be given the results of the tests. There were about 10 groups that had been supported by the Foundation, and after dinner we each got a paper with our results. The next morning we had a meeting to discuss the results. I remember the first person started by saying that he was not interested in detecting antibodies to schistosomiasis anyway, so clearly he had bombed out. I came in as number two, beaten only by the people from the Winches Farm. The reason of course was that we had the purest antigens. But the advantage of the indium slides was that it was simple. I was later invited by the World Health Organization to give a talk in Zurich about schistosomiasis, and they liked my test because it was simple and no tools were necessary. They ended up offering to send me to Brazil where they estimated that 200,000 people suffered from the disease. They did not know that I was a single person working on schistosomiasis at GE, and that I had only worked with schistosomiasis for a couple of weeks.

Trying to commercialize indium slides

My manager at this time, Milan Fiske, wanted me to explore the indium slide invention and we travelled to many different companies to talk about it, such as Kodak, Abbot and Hoffman LaRoche. LaRoche gave me a test; they said they that if I could detect Carsinoembryonic antigen (CEA) at concentrations of 10 nanograms per cubic centimeter, they would consider purchasing the patent. I went down to Hoffman La Roche Laboratory in New Jersey and was successful on my first attempt to fulfill their criterion. In principle you can absorb the antibody on the surface and expose it to the antigen, but when you try that, the antibodies are not very effective. Thus I had to do an inhibition test. This is how the test was carried out: first I deposited the antigen (CEA) on the indium slides in two small dots on each slide as usual. Then I prepared a twofold dilution of antibody and put one indium slide into each vial. When the concentration of antibody is high enough the dots on the indium will be visible because the antibody will react with the antigen on the slide and form a double layer. Using the concentration of antibody where the dots were just visible, we mixed it with a twofold dilution of antigen and incubated it for a certain time. The dots will be visible this time where the concentration of antigen is low enough. By using a second antibody to the first antibody, the test can be made more sensitive. By using a second antibody I satisfied the criterion and Hoffman La Roche agreed to try the test for one year.

I was not too eager to work on practical problems, and therefore it was probably my fault that the test never became developed enough to be used in a hospital or clinic. Another problem was that a new test had recently become available and popular, a radioimmunoassay test, developed by Rosalyn Yalow. She actually received the Nobel Prize for the test. This test was more sensitive than my test and gave quantitative results. That was really not needed for ordinary illnesses, but radioimmunoassay had already secured a place. Using radioactively tagged antibodies made the test much more complicated than the indium slide approach. I actually met Mrs. Yalow a few times. One time we met because of an exhibition at the Science Museum in Chicago. She told me that *Science* had rejected her first paper describing the radioimmunoassay test, and she had saved the rejection slip. She wasted no time telling *Science* that when she received the Nobel Prize. We were both interviewed once on TV. Dr. Yalow surprised me when she said that, as a female scientist, she had to work 16 hours a day. One task she had to do was to make dinner for her husband, who was also a physicist. Maybe it was true that she worked such long hours, but it was not very inspiring for the young kids in the audience. She also believed that a little radiation was good for you, and she should know because she had a Ph.D. in nuclear physics!

After about a year Hoffman La Roche decided that the test was not for them, but Kodak was interested, as they had recently entered biology testing. They were very impressed by the simplicity of the test, but the head of the laboratory said that he did not think Kodak and GE could ever collaborate together. Abbot was the next company; I had travelled with a GE person who was very interested in the test when GE originally tried to sell it. He had left GE to join Abbott and was championing the test there. They were genuinely interested, but unfortunately the former GE person could not pass the medical exam at Abbott and their interest waned.

Chapter 18

Coolidge Fellowship

Right before I received the Nobel Prize I had received the Coolidge prize from GE laboratory management. This award gave me the right to take leave from the laboratory for one year. I was invited by the American Physical Society to spend a month in Japan, and I decided to use part of my Coolidge prize for this. At the time GE had an office in Tokyo. The manager there met Inger and me at the airport, and told us about Japan and Japanese customs. He gave us a slip of paper with our address in Tokyo written in Japanese, which we were supposed to give to the taxi drivers so they could get us home. At the time they really did not have street addresses like we do in the USA. Almost no one spoke English in Japan, and you could not hail a taxi on the street like you did in NY. The taxis simply did not stop for foreigners, because they knew that the passengers could not explain where they wanted to go.

We were very impressed with the bullet trains in Japan, and we travelled in style from Kyushu in the south to Hokkaido in the north. We visited Hiroshima and got a powerful impression of the horror of an atomic bomb. It was a sobering experience; nevertheless I do not think President Truman had a choice at the time. Hopefully the atomic bomb will never be used again. I think that its use has prevented the world from getting into a full scale war for over 70 years. When we visited Kyushu, Inger stayed at the hotel while I went to give a talk at a university followed by lunch at a nice restaurant. I was served some snails in big shells that tasted terribly bitter, but I managed to eat them. At night we attended a nice party complete with Geisha. Inger convinced me to sing a Norwegian song with her to add to the entertainment. She also saw some people at the neighboring table eat something in big shells and asked what they were. Unfortunately, our Japanese host immediately ordered the snails for us, and I had to eat them once more that day.

While in Japan, I went to a meeting held by the Japanese Physical Society, but the talks were in Japanese so I did not get too much out of that. I gave several additional talks at various universities; I remember the one I gave at the University of Tokyo, when I stayed at the GE apartment in the city. A few scientists took me back to the GE address where we stayed, and I asked them how I could get back to the university the next day. They said: "We will pick you up." I told them I could take the subway. They said it was too difficult and dangerous, and that they would pick me up. After some discussion I gave in. Next I asked how my wife was going to get there for dinner in the afternoon and they said: "She can take the subway!"

In Hokkaido I gave a talk and spoke to many scientists in their laboratories. I remember one who talked about "Lat Rivers" that I had difficulty understanding, until it dawned on me that he wanted to say: "rat livers". At least then I knew what we were talking about; for some reason "r" and "l" is confusing for some Japanese people. When I got back to the hotel that night Inger acted a little strange, and said she had bought some ceramics that day. I asked her if I could see them, and she brought out some beautiful vases to show me. I said it would be difficult to take them with us on the plane. But then I accidentally looked behind a chair and lying there were even more ceramics. Inger had thought they were very cheap and she could not resist the temptation to buy them. We had to go out and buy an extra suitcase to get it all home. When we checked in at the Japanese Airline we were grossly overweight, but we did not have to pay extra, as long as we acknowledged that the Japanese Airline gave us special treatment.

We stopped over for a few days in Hawaii to sightsee and have a few days of vacation. I was very happy to get a cheap rental agreement for a car at the airport, but then I found out I could rent a car downtown for a fraction of the price. The reason rentals are so cheap should have been obvious: no matter how hard you try you cannot wear a car out in Hawaii because the weather is always good and after driving 50 miles or so you have covered the whole island. I wanted to try to learn to surf so I rented a surfboard on Waikiki beach for basically the same price as renting a car. It was a calm day, but looking out to the ocean I could see that there were bigger waves maybe a mile or so offshore. So I said to Inger that I would paddle out on my surfboard to investigate, and she could sunbathe on the beach. When I got out there I stood up on the board, but the waves were not powerful enough to really surf. Then I looked down and saw that the sea floor was full of sharp stones. I figured that it could be dangerous to fall off the board and so decided to quit. When I paddled in at first I could not find Inger on the beach; she had been picked up by a young beachcomber looking for adventure. He was visibly disappointed when he saw me coming with a surfboard and understood that Inger and I were a couple. I hope Inger was not disappointed? We played tennis together on the roof of the hotel which was a novelty and had a nice romantic dinner that evening. The next day we drove to the other end of the island where the waves were gigantic. I could not believe the skills of the surfers and how well they handled the waves. While we were lying on the beach Inger said: "Aren't you going to rent a board?" And I answered: "Only if you want to get rid of me!" There are many skills involved in both marriage and surfing!

The remaining part of the Coolidge fellowship year I split between the Salk Institute and the University of San Diego. At the Salk Institute I worked in Jonas Salk's laboratory and at the University of San Diego I worked with Bernd Matthias and his group. To get there we decided to drive across the USA. We started out right after Christmas. John was at the University of Washington in Seattle so it was Inger, me, and the three girls loaded up into the car as we started out on our Californian adventure. I decided that Trine, the youngest, could not bring her large collection of moppets with her. That created some problems, but then Inger protested and loaded the moppets into the car. It turned out that Trine never played with the moppets in California or ever again. Trine's friends in California were more sophisticated than the same age group in New York.

My plan was to start out early in the morning, drive until about 3 o'clock in the afternoon and then find a hotel to settle in for the night. We had planned the trip beforehand with maps from AAA, the Automobile Association of America. If you were a member, the AAA would make you a custom booklet showing turn-by-turn directions. Not exactly Google maps, but it was very helpful. Since we were driving during the winter right after Christmas, we took the southern route across the country. We stayed at Howard Johnson hotels all the way because kids under 18 stayed for free. Also, because the hotels were almost identical; every night when we checked in, we knew exactly where each of us should sleep. The big problem was that the kids were slow in the morning, so we tended to start at noon and drive to 9:00 PM. Also looking at sights along the way took forever. So it became a strenuous trip at least for me, and I would never do it again, except maybe in a sports car with just Inger.

At the University of California San Diego (UCSD), I ran into some problems with patents. The university wanted me to sign a patent agreement with them, but because I was working for GE I could not do that.

The dilemma was solved in a creative way. The UCSD lawyer said: "I will write a letter to the GE lawyer, in a month or two he will answer, and I will then write back one month later, and by that time your time here will have ended!"

In retrospect it was not a good idea to divide my time between two institutions because in this scenario you do not get to feel that you belong in either place. The Salk Institute is located close to UCSD and I could walk from one place to the other. It is located on a cliff overlooking Black Beach and the Pacific Ocean, and is a spectacular location and a beautiful and interesting building. The first time I went to the Salk Institute to meet Jonas Salk I was a little early. Going out to the edge of the cliff that morning, I noticed many old men sitting there with binoculars. When I got into the building and talked to the secretary I remarked that many people must be bird watching at the edge of the cliff. She laughed and said: "Black Beach below the cliff is a nudist beach, so I don't think they are looking at real birds!" Inger and I actually visited Black Beach a couple of times to see what it was like. You see lots of naked men who have various shapes and forms but only a few women. The women though are all beautiful, so Inger got quantity and I got quality!

At the Salk Institute I ended up working with Elsie Ward on a tissue culture project. That was nice because being in biology I think it is important that you work with living things. Elsie Ward had worked with Dr. Salk for a long time and she had helped with the development of the Salk vaccine. She was a nice woman and very organized. For example, when she took a new pencil out of the stockroom, she glued a small piece of paper onto the pencil where she wrote the date, so she could see how long she had had it. I was working with the indium slides at that time, and she did several experiments with cells with me. We discovered that the mammalian cells will not attach at all on top of antibodies attached specifically to antigens on the surface; this meant that cells, for some reason, did not like to attach to the constant regions of the antibodies which were facing upwards from the slide. We published a paper about that, but could not find any explanation for this strange phenomenon. I also helped Jonas Salk's son who worked for his father. He was a medical doctor and did some experiments that used 5 rats. When he described his results to me he used statistics, but I told him, just tell me what happened to the 5 rats! He also told me that he had once thrown 50 pennies on the floor, and they all came up heads. That happens once in 10^{15} trials, i.e. never in anybody's lifetime. I decided to tell Dr. Salk that it was a bad idea to have his son working for him, and went to Dr. Salk's office. On his office wall he had paintings of Françoise Gilot and Paulo Picasso. Françoise Gilot was now Mrs. Salk, but she is best known for having been Picasso's mistress. After some chit chat I said, "An education as a medical doctor does not really prepare you for doing research," and Dr. Salk answered: "I agree with that; the exceptions are my son and me!" Right then I thought it best to drop the subject I had in mind.

At the time Dr. Salk was very famous and Inger and I were invited to dinner with him a few times. Each time the maître d' went on the floor and announced: "We are fortunate to have Dr. Salk with us this evening"; because Dr. Salk had informed them beforehand that he would be at the restaurant. The people in the restaurant then clapped. As famous as Dr. Salk was among regular folks, he was not much liked by the medical profession, probably because he received money from the March of Dimes to start a research laboratory that he named after himself. I was astonished to find out that he was not a member of the National Academy of Sciences, so I tried to remedy that. Bernd Matthias helped me and we talked to many people in the medical field at the university. They all refused to suggest him, so Bernd suggested that we should nominate him in the field of physics. At the time I was not willing to stir up the scientific community. Dr. Matthias on the other hand had no qualms about stirring up the scientific community; he enjoyed that and spent part of his life doing just that. For example, he had his students do paranormal experiments in class. I was a censor on one of his tests where one question was "Describe the universe and mention two examples!" He was always eager for something new or challenging. I actually first met Bernd at a March Meeting of the American Physical Society in a poker game. At these meetings someone always organized a poker game and they played table stakes. I never liked these stakes, but had no choice if I wanted to participate. Table stakes means that you put a certain amount of money in front of you before the cards are dealt. If you use all that money by betting, you are still in the game, but only for that part of the pot. So if there are 10 people or so you may end up with 3–4 pots in a game. It can become very complicated but there is always somebody who keeps track of who is in which of the various pots. Professor Bardeen was often part of these games, and he was very careful, never having more than $10 in front of him. Once I won a pot Bardeen had contributed heavily to and Bernd remarked: "Ivar I bet that is a close as you will ever come to the Nobel Prize money!" That created some laughter because at the time a Nobel Prize was not in the cards for me.

Through my office window at the Salk Institute I could watch the hang gliders taking off from the cliffs at Torrey Pines and it looked like a lot of fun. I asked Bernd about hang gliding and he arranged a lesson for us. He got hold of a teacher, and we went down to some sand dunes in Mexico. Hang gliding is a very easy sport to learn, and the second day I flew maybe a football field in length. It was delightful and I took Inger down to the same place to try it, and she did. However, she stalled the hang glider which made her fall backwards rather hard on the sand. The glider broke and that ended the lesson for that day. Actually, we never went hang gliding again because we learned that our teacher had a glider accident and died; so even though we felt hang gliding was easy to learn, it is also very dangerous.

Surfing, however, is a very difficult sport to do, because you cannot surf on small waves. The beaches in La Jolla are beautiful and ideal for surfing. Very soon after I arrived I asked in a surf shop if they rented surfboards and they asked; "Have you ever surfed?" I said no and they said it is difficult to learn and that I should buy a board. They then sold me the biggest surfboard they had. I tried it for a couple of weeks and did not manage to get anywhere, and gave up. My daughter Guri, however, had several friends who surfed, so she took over the board and learned to surf very well. I took her success as a challenge and during the last few weeks we spent in La Jolla I went to the beach every morning at 7 A.M. to practice surfing. To try and make it easier, I had the surfboard on a leash so I did not have to return to the beach every time I pearled, which is a nice way of saying fell off the board. It is called pearling because when you fall off your surfboard you normally go to the bottom like you were looking for pearls. At first I was alone on the beach which made it easier. But as soon as the surfers started

Ivar hang gliding in Mexico, summer 1973

Inger and a friend trying out the big surfboard

coming, they saw me and wanted to surf next to me. After a while of surfing in their company, I took my board and walked up the beach a few hundred yards and started again alone. They followed me again, thinking the waves were better where I was, but it is more difficult to surf in a crowd.

The beach was beautiful in the mornings. The pelicans flew along the beach, and once in a while they dove for fish. We were warned about great white sharks in the water, but I never saw any. One morning though, I saw three fins between me and the beach, and I thought, I hope they are dolphins. I paddled back to the beach in a hurry, and found that when you swim close to them, the dolphins are huge!

Inger and the kids in California. Inger is in the middle

BACK AT GE

CHAPTER 19

At the GE laboratory I continued to work on biological problems and I enjoyed that. Charlie Bean hired a young biologist from Cobleskill into our branch, a scientist named Charles Keese who had an NSF scholarship and wanted to work at the GE lab for a year. Charlie Bean wanted him to work with me and because he had the right credentials, including an undergrad degree in physics from SUNY, Albany and a Ph.D. in biology from RPI, I decided to take him on. I never regretted this decision and we are still working together after having co-authored around 50 publications. The first paper we published together dealt with a new idea for a radioimmunoassay, where the antibody is tagged after the binding assay is performed rather than before. After the antibody is tagged it is removed with an acid which makes it safer to perform an immunology test because you are not using radiolabeled protein. We were proud of that idea, but the number of citations we have received for that paper is exactly zero. The problem may be that it is a method paper, but we published it in a medical journal. Or heaven forbid, maybe the paper is no good! It is very educational to look in the citation index to see how your various papers have fared, and then try to remember what you felt about them at the time the papers were written. When you are young you are terribly afraid of someone stealing your ideas; when you become more experienced you become more relaxed, and sometimes happy that someone liked your ideas even if they stole them.

In 1979 I arranged for a physicist from the University of Oslo, Dr. Jens Feder, to come to the GE lab for a year. He had his own funding so that did not pose any problems. I knew him well from before; I had also been an examiner when he defended his Ph.D. thesis at the University of Oslo. We worked together on the absorption of ferritin, a protein molecule that carries iron in your blood. At the time we used an electron microscope to view the absorption on a surface, because ferritin is easy to image. Jens also determined the theoretical maximum coverage of protein on a surface, by calculating the random adsorption of fixed size disks on a surface. Jens stayed with me for one year and fitted in nicely with everyone in the laboratory.

Charlie Keese had now gone back to his professorship at Cobleskill University, but spent one day a week at the GE Research Lab. During this time, we were contacted by Dr. Robert Rej from the Wadsworth Center at the NY Health Department who wanted to know if we could detect aspartate aminotransferase isoenzymes with the indium slide technique. It worked well and we published a paper on the technique.

At the time I had become interested in DNA and wanted to try to develop ways to determine the sequence. I knew about Dr. Erwin Muller's technique that allowed him to see single atoms of tungsten. He accomplished this by putting a small and sharp tungsten needle in front of a fluorescent screen in a vacuum. Then he applied a high voltage between the needle and the screen. The hydrogen atoms remaining in the vacuum would ionize on the tungsten needle, follow the field lines and bang into the screen. Using this technique you could achieve almost any magnification. I went to the University of Philadelphia to talk to Muller and to learn more. I later contacted one of his students, John Panitz, and he agreed to work with me on a project. I knew a lot about protein adsorption on metal at this time, and I decided to start with protein molecules rather than DNA which was the ultimate goal. John was located at Sandia Laboratory in Albuquerque, and I needed security clearance to work with John. I dutifully filled out the security forms. The people in the office doubted at first that I was a physicist because I had filled out the questions correctly! This was during the crazy time when the US military trucked nuclear missiles around the USA so the Soviets could not locate them. I suggested in jest that it would be much safer to put the missiles on submarines in Lake Ontario, and the security officer liked my proposal! It was fun to work with John Panitz. We showed that you could indeed deposit protein on a tungsten tip, but it became messy when you pulled them off with an electrical field. However in the meantime, Professor Frederick Sanger came out with a neat method to sequence DNA and I gave up on the original project.

At that point I started thinking about cells instead of DNA. I had a little knowledge about cells from my stint at the Salk Institute. Dr. Keese had lots of experience with cell culture from his thesis at RPI. One of the first experiments we did was to grow cells inside a drop hanging from a pipette. Because protein will denature on the air–water interface, this layer can be strong enough to support the force that the cells apply on the substrate. Coincidentally, we made these observations when NASA was asking GE for research ideas. In fact, every year at the GE R&D Center we get a general request from NASA with a plea to suggest a biological experiment or any experiment that could be performed in space. In general there are no obvious experiments because gravity is a very weak force and it has little effect on the protein machinery in living things. That fact does not seem to bother NASA. They keep sending spiders, grass seeds and other marginally interesting organisms into space hoping to find something significant. So at this time I suggested that we grow cells on the outside of a small droplet of tissue culture media in space. NASA liked the idea, but GE asked me to write a statement that this was a very important experiment that would lead to all sorts of inventions. I did not think it would, so I did not write the statement and the experiment was cancelled.

We were very interested in the strength of protein films and put tissue culture medium on top of silicon oil. When you seed cells out on oil, the protein film that covers the oil can support the traction of some cells, depending on the cell type. We then had the idea of forming small silicon oil droplets for large cell cultures. This also worked well, and we published a paper in *Science*.

Many physicists have tried to perturb cell growth using magnetic fields, and a letter published in *Physical Review Letters* from the Naval Laboratory in Washington claimed that a low frequency magnetic field increased cell growth. So I went to see the experiment for myself. It turned out that none of the scientists involved wanted to talk to me, so I asked to see the experimental setup. What I found was an ordinary cell incubator in which they had set up a big coil providing the magnetic field. When they dialed up the magnet to 20 Hertz, it caused everything to shake. So I readily concluded that it was the stirring of the tissue culture medium that had caused the effect, not the magnetic field. Indeed, to affect living things with magnetic fields you need very high fields which are difficult to achieve in an ordinary laboratory.

Compared to magnetic fields, electric fields are another story, so I designed experiments to test the effect of electricity on cultured cells. I put two gold film electrodes in a petri dish and applied a DC voltage. The cells on one of the electrodes died. That was promising, but the cells died because of chemical reactions going on at the electrodes. We also tried experiments with alternating current and noticed that the resistance we measured did not depend on whether we had cells on the electrodes or not. That was a big puzzle because we could see the cells covering the electrodes, but the resistance hardly changed.

At the time Charlie Bean was experimenting with a Coulter Counter, an instrument that counts particles (or cells) by passing them through small holes; he used it for counting bacteria and viruses rather than cells. To make the small holes he needed, he used nuclepore filters. A nuclepore filter has very small holes created in a plastic sheet by exposing the membrane to nuclear radiation followed by dipping in an acid bath. So Jens put a plastic sheet with only one nuclepore on a metal sheet and managed to put one cell on top of the hole. The cell had little or no effect on the resistance because the resistance of the nuclepore itself was large. We soon recognized that the nuclepore hole was not needed; just a very small electrode was sufficient. In fact, when using large electrodes the resistance of the cells was minor compared to the total resistance, most of which was due to the resistance of the solution that was very high in comparison. The resistance of a small gold electrode in solution is inversely proportional to the area, but the solution resistance is inversely proportional to the radius so by making the electrode small, the actual electrode resistance can be made larger than the solution resistance. We then showed that the trick is to seed out cells on electrodes smaller than 200 micrometers in diameter or about the width of the period at the end of this sentence. With this setup you can easily measure how the cells change both the resistance and the capacitance of the electrodes. Thus ECIS, or Electrical Cell-Surface Impedance as we named it, was born. We were excited about the applications of ECIS and applied for a patent using small electrodes for cell research. Interestingly, the patent office used one of our own patents against us. We had already been issued a patent on the use of a small electrode in immunology. We reasoned at the time that GE would never go into the business of growing cells on small electrodes, and decided not to pursue the patent. Since we were the first to measure the impedance of mammalian cells, this was a mistake, because our new business venture could have made good use of such a patent.

Saudi Arabia

At about this time GE wanted me to go to Saudi Arabia. Inger was not invited but I promised to buy her something. The reason for the trip was to receive a prize that I believe was unique and probably will not be awarded again. I received a beautiful Arabian coffee pot made of silver, which unfortunately developed a crack after a few years, suggesting that the silversmiths cannot be very good in Saudi Arabia. I was the guest of honor on this trip and traveled with a few men from the GE research laboratory. This was before Power Point became the norm for presentations, so I had to carry a slide carousel with fifty scientific slides. At the border we went through customs, and the woman in front of me had about 25 small empty airplane liquor bottles in her bag that she had to throw out. She protested, but that did not work. When my turn came the customs officer asked what kind of slides I had in the carousel and I said: "Scientific slides for a talk." He picked one slide at random from the carousel and looked at it. It was a joke slide I had borrowed from Bob Fleischer, and it showed Dick Tracy getting a bullet through his head. That set off a warning signal and I had to go into a small dark room where we looked at all the slides. There is of course no liquor allowed in Saudi Arabia. When we went to a party where juice was served, you were not supposed to notice that people from the religious police sniffed your drink and checked the hors d'oeuvre for pork.

I gave a nice talk (at least I thought so) where I boasted about science in general and in particular I talked about several scientists: Einstein, Bethe, and Feynman, all of whom I admire. My friends from GE pointed out to me that these scientists were all Jewish and thus probably not very appropriate for the occasion, but I got away with it. Since I grew up in Norway, which was a very homogenous society, I am not accustomed to

Ivar sitting with a few Saudis

Ivar in Riyadh, Saudi Arabia, 1981

classifying people's ethnic origin. I was supposed to be on the radio as well. The interviewer was an English woman, and she needed a ride to the studio, so we picked her up. Later we were scolded by local GE people, because we had transported a woman in a car without her husband. If the Saudis had found out, GE would have been in deep trouble. Inger had heard that gold was cheap in Saudi Arabia and wanted me to buy her some. I had traveler's checks and went to the city square to exchange them. There was a boy of about 16 who stood by a simple wooden desk on wheels with one drawer. He had no difficulty changing my checks, because in his wooden drawer he must have had $50,000 and I needed just $100. In any other capital in the world he would not have lasted 10 minutes, but in Saudi Arabia they cut your arm off if you steal. We had the opportunity to travel a little and drove from Riyadh through the desert to Jeddah for sightseeing. There were a lot of abandoned cars lying along the road almost covered with sand. In Jeddah we had the opportunity to snorkel in the Red Sea, and it was really spectacular. The Red Sea is very rich in different species of strange-looking fish of all colors and I have never seen anything so fantastic since. We were a few hundred meters from land, where a couple of small gunboats were tied to a dock. When we were returning to the shore, one of the small gunboats picked us up, and we had to answer a lot of questions about what we had done. Maybe the officers were bored and needed a break. We had been in their sights for hours.

On the last day in Jeddah before returning to the US, I had an interpreter accompanying me. He was black and an unusually large man so I took a picture of him at the city square. Immediately the religious police appeared and wanted the film in the camera on the reasoning that I could have accidentally caught a Saudi woman on film. I had taken more than 30 pictures on the trip, and they were all on that roll of film, so at first I refused to hand it over to them. I asked my interpreter what would happen if I refused. He said: "Do you see the two real police standing over there with machine pistols? They will call them, you will be taken to the police station, and the police might let you go or they may hold you for two weeks, I just do not know." I reluctantly gave them my camera, and they stripped it of the film.

Feynman

It is almost impossible to write about physics in the USA without mentioning Richard Feynman. I once visited Caltech before I won the Nobel Prize and I gave a talk which he attended. I cannot remember any unusual thing from the talk, but I remember going to his office afterward. We passed by a placard commemorating his 1959 speech: "There is plenty of room at the bottom." Feynman concluded that talk with a challenge; he offered a prize of $1000 to the first individual who could construct an electrical motor only 1/64th of a cubic inch in size. This challenge was achieved a year later by William McLellan. I also recall from that visit that Feynman had inserted himself into a picture of the people who started Stanford University. It will probably surprise no one familiar with Feynman to say that he was not exactly a shy person! On the day of my talk, a student walked me to his office and said, "Professor Feynman — this is our speaker this afternoon, Ivar Giaever, I will leave him here with you," whereupon Feynman ducked behind his desk, feigned terror and said: "With me? All alone in my office?" We had a nice chat about electrons and light passing through one or two slits, and he lectured me about superconductivity. It was a good conversation and a good lead in to my talk which was about type II superconductors, the DC transformer and the magnetic flux quantization. I cannot remember if Feynman asked many questions during the talk.

The second time I met Feynman was when he came to my talk about immunology, probably about 10 years later. The following excerpt is from the memoir of the famous Israeli physicist Yuval Ne'eman, who

I visited in Israel. I did not know that he had written a memoir at the time, but someone sent me this brief story where I play the main part:

> In one sphere at least, Feynman made a deliberate effort to play the role of *enfant terrible*. Combining his love for scientific truth in his hatred of "imposters," he was always willing to play cat-and-mouse games to expose laws and pretense — all of which made him the terror of lecturers who came to Caltech to present their intellectual wares. On my first visit, I gave a series of ten lectures on "group theory," the mathematical apparatus I used, and Feynman interrupted me with questions every 3–4 minutes. I knew the material, however, and in the end it was Feynman who got tired out. On the other hand I remember the visit of a well-known Swiss physicist Johann Jauch, who after half an hour of torture, flung the chalk away, announced he could not go on this way, and walked out the door. In 1976, Werner Heisenberg, one of the leading physicists of the 20th century, made a stop at Caltech as part of a cross-country lecture tour. Since 1950, he had been expounding a certain theory that was generally acknowledged as worthless. Out of respect, however, people usually refrained from arguing with him. His lecture at Caltech too was on this theory. Gell-Man stayed away purposely, but Feynman was there and made his presence felt. At a certain point he got up and shouted: "If that's so, your theory is crap." Mortified, Heisenberg left the hall and according to the distinguished physicist Harald Fritzsch, who was at Caltech at the time and on close terms with him, he never recovered from the shock. He died that same year.
>
> But there was someone who gave Feynman a taste of his own medicine. The Norwegian-American physicist Ivar Giaever once suffered through a lecture with Feynman. Two years later, he came back to Caltech to give another lecture. This time, however,

Giaever not only answered Feynman to the point, but also made him look stupid. Obviously, he had done a good job of preparing ahead, deliberately slipping in remarks to provoke Feynman — who walked straight into his trap. Everyone in the lecture hall could feel how stunned Feynmann was.

I remember Feynman asked several questions during my seminar but I was prepared and did not mind. I remember the audience laughing when Feynman asked how I knew the ellipsometer measured a film of protein on the surface and I said: "The ellipsometer only measures the amount of material on the surface. If Professor Feynman chooses to believe that the protein molecules are on top of each other in a column that is also a valid assumption." But I had not anticipated this question or prepared for anything beforehand so Ne'eman's story is absolutely false! Nonetheless, I take some pride in that at least in some people's eyes I had outsmarted Feynman.

Leaving GE for RPI

In 1988 GE got a new director at the laboratory, Dr. Walter Robb, who was brought in to replace Dr. Roland Schmitt. He had strong opinions about the function of the research laboratory. His mantra was: "From tomorrow on, I want you all to work on *current* GE products and processes." I had already switched to biology, and Charlie Bean and I had started to spend a day a week at RPI, and we were teaching a biophysics course together. At the time GE was not involved in biology research, so in a way I was a *persona non grata*, but I could not be fired because I had, after all, won the Nobel Prize. Charlie of course reminded me that I had said exactly the same thing the day before I received the Prize! I made a deal with Walter Robb to go back and work on superconductivity half the time. It did not take me long to realize that this was not going to be very effective. It is not satisfying to work for an institution where you are not wanted, and so I decided to leave. This was not an easy decision; what made it easier was that my good friend Charles Bean had already left GE to go to RPI. He felt similarly about Dr. Robb's new direction. To me it seemed that Dr. Robb was going to change the direction of the laboratory and the easiest way to do this was to get rid of the most admired people at the laboratory.

By this time Inger was in Norway because her mother was seriously ill. I made it known to the scientific community that I was looking for a job and received some offers. I was tempted to go to the University of California in Davis because they had a medical school and I liked California. But since Inger was away I decided to go to RPI for a short time, because then I would not have to personally move. Dr. Robb also offered me $20,000 if I went to RPI, but when I finally made the decision he cut it to $10,000. Dr. Robb always argued about money and said it was because of my increased pension. I had worked for GE for 29 years, but he offered to carry me for one year so I could get a higher pension. I had no reason to object and accepted the deal. Inger had thought that Jack Welch, the President of GE, would not like that I left, but when I talked to him, he just wished me luck. I also thought I should tell him that with Dr. Robb's decision, the laboratory would not do much real research any more. So I told Dr. Welch that and asked him where the new technology would come from, and Dr. Welch calmly answered: "Ivar, we will buy it!" So the change did not seem to concern him.

I really enjoyed being a professor. You are really on your own, able to do as you please with your time and your work. You are very independent, but must satisfy your teaching obligations, and must apply for grants to support your research. The common thing when you take an academic position is to get a start-up fund to start with, but after that you need grants to support students. I had a grant from the National Foundation for Cancer Research, a small private foundation that I had used to support Charles Keese when he was at GE as a visiting scholar, but now he was on the GE Staff. When I negotiated my new position with RPI I tried to get Charlie to move with me but he declined; he thought the distance to RPI from his home would be too far to travel.

I had my first student at RPI and I decided to continue to work in biology and focus on mammalian cells. It takes a certain amount of time to set up a tissue culture lab, and because you are on your own, you basically have to do the work yourself together with your students. Because Charlie Keese and I had worked together I decided to take him and his wife out for a nice farewell dinner. During the dinner Inger confided to Charlie that I missed him at RPI, and "miraculously" he changed his mind and decided to join me at RPI after all. I was happy about his decision and also that I could support him on my grant.

At most universities in the USA, professors only got paid for 9 months a year, so from 1988 I could take on other work during the summers. Jens Feder wanted me to spend my summers at the University of Oslo and made me an offer. That was good for Inger and me because all our relatives were still in Norway. The big Norwegian oil company STATOIL, which I had dealt with before, set up a program called Vista together with the Norwegian Academy of Science, and I was hired for 3 months a year as a Vista professor. I decided together with Jens to spend it at the physics department in Oslo. There was some misunderstanding to begin with, but finally we settled on a salary of 250,000 Norske Kroner. In Oslo I basically worked with Jens Feder and his friend Torstein Jossang. They were very hardworking and very similar to university professors in the US. Torstein took care of the financial aspects, and Jens dealt with the sciences. I could also have graduate students together with Jens so that was very good. I also had some contact with Morten Laane in biology who was an expert microscopist. He introduced me to a strange organism called Slime Mold that I later had a graduate student work on.

Ironically, I had originally left Norway because I was unable to find any place to live, but by now this had changed. We actually first tried to buy an apartment in Oslo, and made a few offers. We ultimately found an expensive place close to the university where they rented small studio apartments, and that suited us well. One problem was that we only had one room and the bed was a Murphy bed, very narrow and it hung on the wall when not in use. But Inger and I could deal with such a narrow bed because we still liked each other.

So now I had academic jobs in both the US and in Norway.

CHAPTER 20

UNIVERSITY OF OSLO

When I first started to spend summers as a Vista professor in Norway, we rented a small apartment in Oslo. This changed when I took over my parents' cottage in Tonsberg. My father had died many years earlier, but my mother, who was 90+ years old, finally decided to relieve herself of the responsibility of owning the cottage. She really loved the place and wanted to be sure it would be well taken care of. The original plan was that my sister Mette, who is 9 years younger than me, would take over responsibility for the cottage. But Mette was, and still is, not the best when it comes to managing money. I made a deal with my siblings where Inger and I would take over the cottage and I paid each of them their share. The cottage is on the Oslo fjord and, at the time it was "rustic", with an outhouse and no indoor water. We had a carpenter add an indoor toilet, a sauna, and an additional bedroom. We have spent every summer there since 1992 and we love the place.

Our beloved cottage in the Oslo Fjord

My always helpful sister, Mette, who paints and loves flowers

Ivar with brother John, at an island in the Oslo fjord

After moving to Tonsberg, I still worked at the University of Oslo during the summers and commuted to the cottage. The trip to University of Oslo was little more than one hour by train and then half an hour by streetcar, so it was okay to commute for a couple of months. There were several people on the train who made the same trip and they typically slept all the way. In the beginning of the summer I worked on my laptop and felt very superior to my fellow passengers because I was being so productive. But everyone gets tired of commuting after a while, and by the end of the summer I mostly slept on the train as well!

I had graduate students in my lab in Norway. It is difficult to have students at a university when you are away 9 months a year. At the time we did not have Skype or other video calling technology to check in every day. Some students are okay with little supervision, but others are not and their work suffers. For example, I started a program in Oslo to look at DNA which was a lot of fun and quite exciting, but unfortunately went nowhere. In Norway Professor Uglestad developed superparamagnetic, micron-sized (one millionth of a meter) plastic spheres which were very uniform in size. I had the idea to hang a small Ugelstad sphere on a short thread of DNA and then observe the random motion of the spheres. This was a difficult problem for me, but fortunately I got help from my daughter Guri who knew how to purify and work with DNA. By using common molecular biological tricks we managed to hang the small spheres using threads of DNA and take pictures of the spheres to observe their random motion. By exposing the sphere to enzymes that cut the DNA the amplitudes of the oscillations increased. We figured out that this was because the spheres were originally attached to many DNA threads and so they were not free to move about a lot. As we cut the threads, the movement of the spheres increased until it stopped altogether once all the threads were cut and the spheres fell off.

I was very proud of this project and I had a lot of help from Jens. The problem was a student named Tom who probably felt uncomfortable

Ivar showing off his skill playing cards for students at NTH

Ivar with professors and students from Oslo at his house in Schenectady

because he was basically an applied physicist and did not fully understand the problem. We could get the system to work when I visited Oslo, but when I was not there, Tom could not get it to work. I tried to tell Tom that we were not making a garage door opener that has to work every time, and he just needed to keep trying and trying, because once he got it working the data would be clear. But Tom was not ready for such a project. One day we were called in to the office of the physics department chairman who wanted to know how Tom's thesis was going. They asked him what he had learned so far and Tom said: "Nothing." Then they asked whether he had learned what did not work and Tom said: "No." This in essence was the problem: exploratory, experimental science is not for Tom. He obviously did not want to continue for his Ph.D. and reluctantly I dropped the project. Some people have "good hands" in the lab and others do not. But if you are not passionate about the science you are better off finding a different occupation.

My experience as a Vista professor

When I had been a Vista professor for about 12 years I was suddenly asked to justify myself. Here is what I sent to STATOIL:

> I have been a Vista professor for 12 years. My biggest surprise is that I have never been asked to give advice about research, education or technology by anyone in the government. I am surprised that I never have been asked by a Norwegian firm to give advice about science.
>
> Everybody knows that all the universities are free in Norway. Unfortunately this is not true in the USA. It is strange for us in the US that young people in Norway are not streaming to the universities. I think the reason for this is that educated people make too little money; after taxes there is little difference between a bus driver and a professor. Another reason is that the unemployment rate is very low. The universities operate in a fashion that would have been good 50 years ago. Professors who teach a course do not set the grades for that course; another professor does. There is no required homework and the students do not have to attend the classes. There are no tests during the year, only a final examination. Both students and professors seem to like this. To use an analogy, the Norwegian Railways (NSB) have had many problems these last few years. The reason is that they do not like to change. Oslo University is like NSB. But the people employed at Oslo University are arguably smarter, and therefore it is even more difficult to change the system. I have tried. When I teach at Rensselaer Polytechnic Institute, I give homework every week. It is unthinkable for me that people can learn mathematics or quantum mechanics without homework. When I tell my students in the USA that there is no homework at UIO they think I am joking. Rather than have another professor

grade my students, at Rensselaer Polytechnic Institute I grade as follows: to be in class 10%, homework 20%, two tests 20% each and examination 30%. It is obvious that there are high quality students in Norway, but the average university student learns much more in the USA than in Norway.

In Norway as well as in the US, professors need grants to fund their research. In general in Norway the money is distributed more or less equally between the professors by the government, with little attention to research quality. In sport, Norwegians support the best athletes and not all the average athletes. In science as in sport, average is not good enough, and so it is strange that the government treats science differently. At Rensselaer about 50% of a laboratory's funding comes from corporations like IBM and GE. This makes sense from the corporation's point-of-view; by doing this they can find out how good students are before they hire someone. I like the fact that STATOIL supports Vista. But I think STATOIL could do a much better job by finding out how their money is used, and learn more about the quality of the students they support. I worked for GE for 30 years, and the most important job I did was to try to hire people better than myself. In the USA it is rather easy to get rid of a person who is not performing well. In Norway it is next to impossible to fire somebody. Therefore by hiring a lazy or incompetent person, 40 years' worth of salary out go the window. It is therefore incomprehensible to me that it is relatively easy to hire someone at the university or in the industry, but rather difficult to receive money for needed scientific instruments.

These are a few thoughts I have had since I was employed by Vista. I have tried to do good science and have had the excitement of having had a few good students in Oslo and the despair of having had an equal number of bad ones.

In the year 2001 (which is typical) I have participated in conferences and given talks at the following places:

April 20–25 Geilo, Norway
June 5–8 University of Bergen, Haukeland Hospital
June 14–17 Oslo Impedance conference
June 24–29 Lindau Germany, Nobel Prize meeting.
June 29–July 3 Stuttgart, Germany, American sector
Sept 18 Oslo, UIO
Sept 22 Bergen, European Young Scientists
Oct 29 Norwegian Embassy Tokyo, Japan
Nov 14 Oslo, Norwegian Museum of Cultural History
Dec 4–6 Oslo, UiO

APPLIED BIOPHYSICS

CHAPTER 21

Since Charlie and I worked in biophysics and studied cellular behavior, we did not fall neatly into a single science category and as a result we had difficulty funding our research. Most agencies in the US pay lip service to how important interdisciplinary research is, but few of them back it up with funds. We applied for grants at both the National Science Foundation (NSF) and the National Institutes of Health (NIH) with little luck. Then Charlie heard about the Small Business Innovating Research (SBIR) program and suggested that we apply to them for funding. Since our measurements of impedance caused by mammalian cells cultured on small electrodes were new, we decided to go with a grant proposal in this area or as we called it Electrical Cell-substrate Impedance Sensing (ECIS). The business would be to teach researchers that by monitoring cells grown on metal surfaces, it is possible to learn a lot about the cells. For example we can see how the cells respond to various drugs, insertion of DNA, or various physical parameters such as temperature. Our customers would then be professors, medical doctors, and drug companies.

The SBIR program was established to help small businesses conduct research, which is admirable. But professors do not like the program because the program requires every grant-awarding government agency to allocate 2.5% of their budgets to the SBIR program which means that they get 2.5% less. I was very reluctant to start a business as I had too much going on already. But since Charlie relied upon soft money (i.e. grants) for his livelihood, I reluctantly agreed. To qualify for SBIR funding you must be able to demonstrate that your research has the potential to be commercialized, so your idea has to be both scientifically valid and hopefully also have commercial interest. There are other rules: at least one person has to work for the business more than 51% of the time; the business must be a domestic US company; and the business cannot have more than 500 employees or be dominant in their field. In Norway even 50 people is a big business and although we were the only ones measuring impedance of mammalian cells, I supposed that two people cannot "dominate a field". So we applied and received a Phase I grant. The SBIR grants are given out in three phases: Phase I funds of up to $150,000 for a six month feasibility study to try out your idea. If you are successful in Phase I, you can apply for $1,000,000 in a Phase II grant for two years to take the idea to the edge of commercialization. In the last phase, Phase III, you get no more money but are expected to commercialize your idea.

To start a business in the US you need a lawyer because the government wants to know where they can go and collect their taxes. Our lawyer told us to pick a name for our business. I remembered a story where Hewlett and Packard had flipped a coin to settle on whose name should come first, Hewlett-Packard or Packard-Hewlett, and I suggested to Charlie that we should call the business Giaever-Keese. For some reason he thought Keese & Giaever sounded better. So after a lengthy discussion we settled on Applied BioPhysics, Inc. and we paid the lawyer $ 562.86 for his services. Before starting the company, we applied for a regular research grant from the NIH which was rejected. We submitted basically the same application to the NIH as an SBIR, and they approved it. They said it one of the best applications they had seen. Since the science was identical, there must have been a different set of reviewers, maybe medical doctors for the first submission and engineers for the second? Or maybe the criteria for good science and good products are different?

Once we were incorporated we needed a place to set up shop. RPI allowed us to use my laboratory for a short time while we investigated new sites. Once possibility was on the RPI campus; they had an incubator center where small businesses can get started. The incubator had Xerox machines, a secretary, meeting rooms, internet connections and a coffee machine that many small companies shared. Even though it was not cheap, it was safe so we decided to move in. On a bitterly cold New Year's Eve day in 1993 we rented a truck, loaded up some used furniture and equipment, and moved in. Applied BioPhysics was on its way.

To run a business, even one as small as ours, you need an accountant, because as far as the IRS is concerned, there is very little difference between a start-up and huge corporations like GE or IBM. Right after we got started, we missed a NY State tax deadline and got fined. So we hired a part-time accountant and she made NY State forgive us. Noreen told them that we were a new company and had no idea what we were doing, which was accurate. Because our Phase I grant qualified us for the more lucrative Phase II grant, we applied, got it and had some financial breathing room. We also sold our first instrument to a medical school in the Midwest. The sale was by word of mouth, because we had never actually advertised anywhere except by publishing our research, and by giving talks and posters at scientific conferences. We were certainly not properly prepared for this sale, and the instrument software was not "user-friendly". Both Charlie and I went to install that first commercial machine, and I spent 12 hours programming to get the system up and running. When we left the medical school we were very proud of the instrument: we had sold something, were really in business, and had made a small profit.

Then, as now, when you buy the ECIS system you also must buy the consumables, which are the electrodes that the cells grow on. This is like the razor and razor blade model where you sell the razor, but rely on the blade sales to provide a steady flow of income. Making the electrodes from scratch was very labor intensive and so, to free up our time to continue research and grow the company, we hired our first technician, Nurayan. He was an Indian student who had flunked chemistry, and he did not dare to tell his father that he had failed. So he had to spend an extra year at the college. He turned out to be a hard worker, and a careful experimenter. We tried to hire him permanently, but he left to pursue computer science.

It is difficult to have both students in your lab and a business at the same time; you have to make sure that you do not exploit your students for your business. Since ECIS was a new area this was not a big problem because there were lots of good research projects to choose from that could benefit the students. As a way to keep things interesting at work, Nurayan and I played chess, at the rate of one move a day. He was a good chess player so it was entertaining. A few years later however, I had a student, Michael, who was originally from Jamaica. He was a very nice person, but could not pass the difficult physics qualifying examination, and had to leave with a master's degree. He also wanted to play chess and so I asked him whether he knew chess, and he said yes. So I said: "Michael why don't you start". When I looked at the chess game he had moved the black king pawn two steps forward! So I said; "Michael, I know you are black, but in chess white always go first!" This was the problem; Michael always thought he knew more than he actually did. I also had a few Chinese students who were in general quite knowledgeable. However, as part of their culture, they had a strong desire to please their professors. As a professor I want my students to surprise me with original ideas and not spend their time figuring out how to please me. This can be a difficult concept to teach, especially if your students do not have a tradition of challenging you and your ideas. For example, when I start

John, Anne, Guri and Trine at my retirement party from RPI, 2004

a new experiment I usually have some vague idea how to begin, but as soon as I start, I recognize the flaws and start over with a new approach. This process repeats itself several times before I settle on a final method. This is really the nature of experimental research. When I start students on something I expect them go through the same process, but the Chinese students cannot believe that their professors can be wrong and stay with the original approach. I am probably not an easy professor to work for as I change my mind a lot.

I retired from RPI in 2004 when I was 75 years old. You are not required to retire anymore in USA and can stay on as long as you can do your job. Shirley Jackson, the president of RPI, had a party for me at the campus, and in addition we had a party at a nice restaurant in Schenectady. All our children, their spouses, and many colleagues attended; it was very nice. It was the end of my academic career, so now I could concentrate on my firm.

KOREA

CHAPTER 22

On my first visit to South Korea I was invited by Edward de Bono. He is a medical doctor, but his mission in life was to teach people how to think. He invited twelve Nobel laureates and their wives for a 10-day trip, providing us with $5000 each and first class airfare. I do not know who paid for the trip, but I suspect it was the government of Korea, as we got to meet the President. I was not going to accept but Dr. Keese talked me into it. It was my first junket trip, and I had an interesting time; not least because I got to know the 11 other laureates. De Bono was also an interesting person; he said that we teach people everything now, even sex, but we do not teach people how to think. He has written several books dealing with how to think, and he has several interesting ideas. After our return he asked me to write an introduction to one of his books, "I am right, you are wrong."

Seoul makes New York City look small, 2009

I have visited Korea several times since then and I am a member of the National Academy of Korea. But in 2009 I was invited to spend at least one month in South Korea, and both Inger and I enjoyed our time there very much. It was a program the South Korean Government had organized to try to increase the chances of a Korean scientist winning a Nobel Prize. I am very impressed by South Korea and I cannot quite understand the fervent desire of both China and South Korea to receive a Nobel Prize. I have been told that in 1953 when the Korean War ended, Korea was broke and had no industry. Today Seoul makes New York City look small; Korea builds more ships than China; Korea makes better cars than Japan; and Samsung competes successfully with Apple. They have managed the impossible, so why is the Nobel Prize such a desirable goal? What they have achieved is so much better. I spent one month in Korea for three consecutive years at Kyungwon University, a university operated and owned by Doctor Lee Gil-Ya. She is a medical doctor and a remarkable entrepreneur. I spent my time in the Nano-Bio Department, and after I left the university it merged with a medical school she had also started called Gachon. They have about 14,000 students and maybe 800 staff, so it is a good sized university. On one of the first days I was there, we discussed retirement age, maybe because I was 80 years old. I told them that there was no longer a fixed retirement age in the US. I was told that at their university there was a strict retirement age of 65 years old. I objected and said, "You told me that the president was past 70 years?" "She owns the university!" was the laconic answer I got, and that explained it!

The Koreans paid well and Inger and I had a big room at a hotel for teachers that was run by the government I believe. I also had a car and a driver to get back and forth to the university. We ate breakfast and dinner at the hotel. I in particular enjoyed the breakfast because of the fresh salad and spicy food I could choose. Dinner was also good, and it was

Korean students in my office

served in the same room as breakfast. In addition they had a large dining room with fancy food, with maybe 100 foods to choose from. I tried them all, and live snails were the most difficult to eat for me. It was a surprise to us that few Koreans spoke English and even at our hotel the staff were not very fluent. We became good friends with many of the hotel employees even though we could not speak much with them. In particular we liked the doorman who was a very nice person and knew a few English phrases. We also became good friends with the hotel director who sometimes sent different cookies to our room. Inger asked him once for bicycles, and he nicely provided us with a used bike each. Right near the hotel was a beautiful bicycle path that followed a small river. In one direction we could reach downtown Seoul, in the other direction we came to a small, nice village or town. The path was beautiful and it even had toilets along the road. We really made good use of the bicycles when we were there.

Seoul has a modern and wonderful underground system that is easy to use. Every station has clean toilets which are free, and the subways are free for retired people as are the entrances to the national museums. When we tried to buy tickets for the subway from the automats people who could manage a little English always offered to help. The government has increased English in schools so the younger people will soon know English very well.

We had some big stores quite close to where we lived; among other things they had a Costco store. At the time I did not know that it really is an American store. We went there once trying to buy a bottle of gin, because we could not find any store carrying liquor. At Costco they had both gin and whiskey and we tried to buy a bottle of gin. But to buy anything at Costco you have to be a member, and we did not want to pay for that. But then we discovered a 20% alcoholic drink in a small local Korean store, that was quite OK, to drink, and started sharing a bottle every day. I asked someone why they sold these drinks so cheaply, and was told that many Koreans had worked very hard to build Korea into a modern nation, and this was one of their rewards.

At the university they were supposed to teach in English, but I had difficulty understanding many of the professors. In the Nano-Bio Department where I was, most of the graduate students were from a particular province in India and had a peculiar English pronunciation. During the time I was there I gave a couple of lectures every week, and I mostly talked about biophysics. I never knew what the students understood from what I said, and in particular what the undergraduate students understood. Koreans are exceedingly polite and I think they never argue with their teachers or with their parents. At the university whenever I saw a student far down a corridor the student would always turn towards me and bow.

A luncheon feast with students and their professor in Korea

Inger and I once went to Ulsam where I got an honorary doctorate degree. The president took me to Hyundai Heavy Industries where I also gave a talk. I believe Ulsam University was started by Hyundai; in Korea it is quite common for an industry to found a university. I was told, for example, that Pohang University was started by the steel industry. We were invited to Hyundai Heavy Industry for lunch, and although I am not much for fancy food, it was the best lunch I have ever eaten. The vice-president of Hyundai and the president of the university were present and the vice-president was very entertaining. He told us how he had visited the model ship building tank at NTH in Trondheim, my alma mater. We also hit it off because he was an eager marathon runner and I could share with him my New York City Marathon experiences, a race that I had run 3 times. In my first NYC marathon, my daughter Anne asked me to run with her. I still remember the long wooden troughs set up at the start for men to pee in. It was a river of urine. I also told the Hyundai VP about my first marathon, in which I wore a GE T-shirt. I got tired at the 17 mile mark and started to walk. Then I heard a voice from the crowd yelling, "My GE refrigerator isn't running either!" So I waved and started to run again. I do not remember, but I think that the director had more than 25 marathons under his belt.

I was interviewed by a journalist after that delicious lunch and he asked me why Korea did not have any Nobel laureates. I hesitated at first but decided to tell him what I thought and said simply: "You are too polite!" He was puzzled. I went on to say, consider the Jewish people; they are 0.2% of the world's population but are involved in roughly 20% of all Nobel Prizes. The Jewish people I know love to argue and challenge each other and I believe that is part of the reason they are successful. I also said that I loved that the Korean students were polite towards me, but I really wanted the students to surprise me and to not believe everything I say. This discussion created quite a stir in Korea when it reached the national newspapers. I was also thinking about the Korean airliner that crashed in

Conference with junior students in Korea

New York because the second pilot was too scared to tell the first pilot that they were out of fuel. Fortunately, I did not mention that event. Inger says that sometimes I talk too much, particularly when I drink.

In South Korea we once took a trip quite close to the North Korean border, to give a talk at the American High School. From there we could look over to North Korea. It is a shame that Korea is divided, and as far as I know many North Koreans are suffering and starving. People in the south would like to help their relatives but North Korean politicians blow hot and cold; sometimes visits are allowed while other times they are denied. I think back to the time when I was in East Germany in 1987, and asked many people in both East and West Germany about the chance of a unification. People typically said things like "never" or maybe in a hundred years. Yet it actually happened only three years later, so maybe we can hope for a miracle in Korea as well?

CHAPTER 23

LINDAU

Every year there is a meeting of Nobel Prize science laureates in Lindau, Germany. Since there are three prizes, medicine, chemistry and physics, the physics meeting used to happen every three years. But they have recently introduced a common meeting for all the prizes so now the physics meeting is every 4 years. In my opinion these meetings are very valuable for the students and also for the Nobel laureates. In Lindau you have an opportunity to meet Nobel laureates whom you would rarely see at an ordinary meeting. The first three meetings I attended I had the chance to chat with Dr. Paul Dirac. Dr. Dirac is arguably one of the most admired physicists that has ever lived. He was not easy to chat with, but his wife was and she helped keep the conversations going. Interestingly, she was the sister of the Nobel laureate Eugene Wigner. There is a story about Dr. Dirac that he tended to introduce his wife as "Professor Wigner's sister!"

An American theoretical physicist once told me that he had read Dirac's book on quantum mechanics at least twice, and then he went to

Ivar giving Maikafer speech in Lindau 1976 in front of physics Nobel Laureates

Cambridge to attend Dirac's lectures. At the lecture, he described how Dirac came into the lecture hall with his book, which he opened and proceeded to start reading aloud. After three such sessions my friend told me he went to talk to Professor Dirac, and asked him if he could not perhaps deliver an ordinary lecture, to which Professor Dirac answered: "I have thought carefully about this, and written down the best way to describe quantum mechanics, so I cannot do better than read aloud!"

There are speculations that Professor Dirac "suffered" from Asperger's syndrome. A second story illustrates that this might be true. During the question period after one of his talks somebody asked "Professor Dirac, I could not understand you when you said such and such." After a long pause the chairman said, "Professor Dirac are you going to answer the question?" To which Professor Dirac answered: "It was a statement, not a question!"

Later, after he had moved from Cambridge to Tallahassee I ran into Professor Dirac at a conference, where we were both giving talks. We sat on some stairs in the auditorium with our overhead transparencies awaiting our turn to speak. A few minutes before it was Dirac's turn, his overheads fell to the floor, and as I was helping him pick them up, I noticed that Dirac looked quite shaken. I was a little afraid that Dirac would make a fool of himself, but he gave a marvelous talk. I still remember the opening. Dirac showed a photograph of a man and said: "You don't know this man, but you will!" Dirac was referring to a scientist who studied nuclear tracks in plastic sheets. He claimed he could recognize nuclear tracks that had to come from the element 116, which was thought to be quite stable. As it turns out the scientist was wrong, so although I remember the talk, I no longer remember the guy's name!

In 1982, the Nobel laureate Professor Kapitsa from the Soviet Union was allowed to visit Europe. He had received the Nobel Prize in 1978 for the discovery of superfluid helium. He had known Professor Dirac since before the Second World War when Kapitsa spent 4 years in Cambridge from 1930–34. After the war, he was restricted from travel outside the Soviet Union like most Soviet scientists at that time. They had lots to talk about. Professor Kapitsa was hard of hearing and therefore he spoke very loudly because he had difficulty hearing his own voice as well: therefore anybody who wanted to listen could follow their fascinating conversations.

At the Lindau meetings, the organizers take great pride in making you feel very important. I know deep down that I am not, but nevertheless this special treatment gives my ego a boost. Since there is no airport in Lindau most people fly to either Stuttgart or Zurich. The organizers pick you up at the airport in a brand new car and drive you to Lindau. You get no money for participating, but you do get free travel and get to stay at a first class hotel by Lake Constance. If you go downtown in Lindau and need a taxi, you just sign your name and never have to actually pay for the taxi. If you give a talk (and that is of course what Nobel laureates really want to do) your wife can join you for free. In addition to a talk you also have a discussion with the students in an informal session for

3 hours or so. You have given your talk previously and now the students can question you. Or better yet, ask you any question at all. I think it must be very valuable or at least interesting for the students to talk with us and hopefully discover that we are really ordinary people.

The first time I ever saw a Nobel laureate in person was when I was studying for my Ph.D. at RPI. Dr. Brattain, the co-inventor of the transistor, gave a talk. I remember being very curious about how he would function and act. I was impressed by his talk which was about semiconductors. According to the Lindau Website I have participated in 17 Lindau Nobel meetings: 12 in physics, 4 in interdisciplinary meetings and 1 meeting in physiology and medicine. Hopefully I will be able to continue a couple more years, and meet with many more students.

On the last day of the Lindau meeting we go together with all the students on a big boat to Mainau, an island in Lake Constance. On the boat the students again have a great opportunity to talk informally with the Nobel laureates. Also on the boat they have small expositions of several student projects which are interesting to look at. I enjoy talking to the students and make myself available for discussions. Once a German student asked me if I knew John Giaever who was his roommate during college in Seattle. When I said I am his father, at first he did not believe me because he had lived with John for two years and John had never mentioned the Nobel Prize.

The best way to describe Mainau is as the "Disneyland" of flowers. Count Bernadotte, one of the three founders of the Lindau Meetings, was the brother of the late Swedish king and an uncle of the present king. He told me that, as a young man, he was not very cooperative and the royal family did not really know how to treat him and what to do with him. His grandmother owned Mainau and on that island there was a church and a castle. The whole island was overgrown and wild. So Count Bernadotte had the idea that the mild climate of Mainau would be perfect for a flower garden. He hired 300 gardeners and the results were spectacular. More than a million people visit Mainau every year, and it has now become a foundation.

One of the reasons for the origins of the Lindau meetings was that Germany was excluded from science after they lost the Second World War. In the early years of the meeting, most attendees were German students, but it rapidly grew in popularity and now about 80 to 90 nations participate. Even though most of the students' expenses are covered and students come from all over (for example India and Pakistan) very few come from Norway. I looked into this puzzle once, and found that although the Norwegian universities are invited, they do not disseminate the knowledge very well to the students. There are, however, many students from the rest of Scandinavia.

Ramsey, Bloembergen, Glaser, Ivar and Mossbauer at a lecture in Lindau

When I go to Lindau to give a talk to the 400 or so students that attend, I try to make it very simple as there are all sorts of students with different abilities and cultural backgrounds. In 2012 I gave a talk about "The Strange Case of Global Warming" which was very popular and the Lindau website first said it was visited more than 4000 times. But since the next talk had maybe 200 visits, they changed the category to Most Visited without giving any numbers. I guess it became embarrassing for some people, as most people believe global warming is a fact, even though the highest temperature measured over land was back in 1998.

It is, of course, a controvertible subject and in my opinion it has religious or political overtones. The American Physical Society states:

"The evidence is incontrovertible: Global warming is occurring. If no mitigating actions are taken, significant disruptions in the Earth's physical and ecological systems, social systems, security and human health are likely to occur. We must reduce emissions of greenhouse gases beginning now."

I resigned from the society because of this statement. In my view nothing in science is incontrovertible. The American Physical Society is willing to discuss the mass of a proton, the speed of light or multiple universes, but global warming is incontrovertible? The average temperature of the whole Earth for a whole year is rather meaningless and cannot be measured in my opinion. But people do and the increase in average temperature is 0.8 degrees Celsius in one hundred years! To me that means that the temperature has been amazingly stable! And conditions on the planet have improved tremendously in the same hundred years, so why is the statement from the American Physical Society so negative? The fact is that no one knows what the best temperature for the Earth is, but that CO_2 is beneficial for plant growth is incontrovertible! And CO_2 has much less effect on the global temperature than most people think, and that is the reason for the failure of computer models to predict future temperatures.

Leo, Ivar and Brian met at Lindau for the second time.

The reason I got involved with global warming is that the USA and other countries spend an incredible amount of money trying to fix a problem that basically does not exist, rather than spending the money on places where people and children actually starve. Cheap energy has been the key to the incredible progress in living standards in my lifetime, and solar energy and windmills threaten to slow this down.

I only have good things to say about the Lindau meetings, they are a good idea and are carried out well. Last year I went to the Indian city, Allahabad, where they tried to copy the Lindau model by inviting several Nobel laureates. They did a good job, but Allahabad is a very difficult place to get to. We had to fly first to Delhi, then take another plane to Varanasi, and finally drive to Allahabad. Google Maps says the drive is 2 hours, but in reality it was more like 5 hours. The traffic in India has to be seen to be believed — it is incredible. No one follows the rules, and you have to pass lorries all the time while somehow avoiding oncoming traffic. Although we had a very nice and skilled driver, the drive was nerve-wracking. The accommodations were very nice and the campus very clean and orderly. In India in general there is a lot of garbage everywhere, yet somehow the residents don't notice it. At the Allahabad meeting they had a giant tent set up on the campus, where the Nobel lectures were given. The students however sat far away from the speaker. This was odd because the lectures are really supposed to be for them. They had difficulty both seeing and hearing the lectures, I believe. The important people, like professors and visitors, have seats up front and have much better access to the lectures. As in Lindau, the Nobel laureates tend to speak to each other, and do very little to try to be understood by the students. The interaction with the students takes place during coffee breaks, where we were surrounded by students who wanted a selfie with us and maybe an autograph.

SPORTS AS AN ADULT

CHAPTER 24

Skiing has been an important part of my life, both downhill and cross country. Inger and I grew up cross-country skiing in Norway. For example, to get to school you had to have cross country skis. At Easter time in Norway many people take a week off from work to ski in the mountains. At that time of year the sun starts getting warm, and people like to get a tan in the mountain air. Inger and I had enjoyed such trips as well, and we had a very nice time as teenagers when we rented a cottage one Easter close to Skogshorn with several of our friends.

I remember quite clearly that Inger and I took some very long cross-country ski trips while our friends sunned themselves close to the cottage. On one of the trips Inger was unlucky and broke the tip of her ski. She started to cry because it was very difficult to ski with a broken ski tip. I saved the day because I had a reserve tip in my backpack and could temporarily fix her ski long enough for her to complete the trip. I think that maybe one reason she agreed to marry me was because I am good

Inger and Ivar at Powder Mountain, 1995

at fixing things. Or maybe it was because we actually shared a bed at the cottage? Skiing was not popular in the USA. When we arrived in the winter of 1955 there were no skis to buy, so we had our cross-country skis shipped from Norway. I remember that it cost more to ship the skis from New York to Schenectady than from Oslo to New York. The first time we went to a ski area in the US was at Mount Snow in Vermont. To our surprise, when we arrived at the base of the lifts, we discovered that unlike Norway, you had to pay for tickets to get to the top of the mountain. In typical Norwegian fashion, we decided to walk up. When the ski patrol saw us, they tried to catch us but they were no match for us with their heavy slalom skis. Nonetheless, we soon recognized that it was too dangerous to walk up the mountains with other skiers racing downhill. Our good friends had bought downhill skis and wanted to ski with us so we all went to a ski area in Stowe, Vermont. It was bitterly cold on the ski lifts, with wind chill factors in the negative. The lift operators passed out big ponchos to wear so we would not freeze on the ride up. Inger and I still had our cross country skis and bindings, and it was not easy to negotiate the steep hills and icy slopes at Stowe. So back in Niskayuna, we bought downhill skis. I had to argue with the salesman at the store. He wanted to sell us long skis, but with our experience from Norway we knew short skis would be easier to handle.

We took the kids skiing a lot every winter. Sometimes we went to small local areas; other times we spent a few days at big areas like Mont Tremblant in Quebec, where we rented a hut. We were pretty serious about getting as much skiing in as possible. Normally when we skied we brought our own lunch, and we always skied during the normal lunch hours, because the lift lines were shorter. We normally skied with our children, and I'm not sure how much they enjoyed it. Because skiing was expensive, one goal was to make as many runs as possible in a day. One infamous family story took place on a trip to Mont Tremblant. When we came in to say good night to the kids, I found Trine had gone to bed with her ski boots on. When I asked her why, she said: "I wanted to save you some time in the morning Dad, when you help me put my boots on!"

One year we went to Aspen and recognized how much better the skiing was in the west. Eventually we got together with some of our good Norwegian friends from NTH. My best friend from that time, Magne, organized the group with Magne, Kjell, Arnold, and Peder who all went to NTH at the same time and were lifelong friends. Of course the wives also joined. At first we went to several different ski areas, like Jackson Hole or Lake Thao or Alta, but then we decided to buy two timeshare units at Powder Mountain close to Ogden, Utah. Erik, who had also gone to NTH a couple of years before us, also joined the group. He was a good addition because he was a good singer and had a guitar. Arnold scouted out the timesharing places, first looking at Alta, but then found out that Powder Mountain was much cheaper. On our first trip there I was kind of disappointed because the skiing was too easy, but as we grew older that objection disappeared. The best thing with Powder Mountain was the large area of untracked snow, with wonderful powder skiing off the beaten track. One problem with the Powder Mountain area was that it was difficult to get to. The condominiums were a mile or so from the ski area and were located on the very top of the mountain at about 9000 feet. Nobody wanted to sleep on the second floor because sometimes it took time to adjust to the altitudes! When skiing with my Norwegian friends I always thought of Valhalla, the Viking version of heaven. At Valhalla you fought all day and at night you went back to drink and brag about your achievements. Although we did not fight, we skied all day, and certainly drank and bragged at night, and we had a very good time for many years.

Part of our Norwegian ski group. Inger at left, Ivar at right

When I was 83 years old I took a skiing shortcut to our condominium and slid down a snow bank at the parking lot. My skis caught on a wooden pole in the snow left for the snowplow that cleared the parking lot. I fell forward six feet onto the sheer ice below. It hurt, but I tried to systematically move my legs and arms, and decided nothing was broken. Then I tried to get up but could not. My friends saw me from the window of the condo and came out to help. I needed the help of two people to get up and it hurt like hell. I insisted that I was only bruised, but Erik drove me to a local hospital in Ogden, while Peder held on to me in the car. A couple of strong hospital workers got me onto a stretcher and then onto a gurney. I was lying on my back and I can still remember the ceiling light in the hospital corridors. They lifted me onto an x-ray table and started to work on me. I had to keep my toes together, and it hurt a lot. I had broken the neck of my femur. They said I was lucky because the best surgeon was on call. He came to speak to me that evening, and seemed competent. I asked if it was a simple fracture, and he said "they are all complicated", so I knew it was a difficult fracture. My son John had already talked to him on the phone and suggested that I should go to a hospital in New York or Los Angeles. But the doctor responded that he had done more operations than my son could count. John answered "I can count pretty far". They must have come to an agreement because I was operated on the next morning.

I spent only three days in the hospital, and was then shipped off to a rehabilitation center in Utah. It was a very good place and the service I received was wonderful. People in Utah, Mormons or not, were very caring. Inger was allowed to sleep in my room and was a great help to me. They offered her a private place but Inger wanted to stay in the same room to give me comfort. I had rehabilitation training every day by very competent and caring people. After completing the rehab they drove us to the airport to fly home. Everything by the way was paid for by Medicare.

Next year we went back to ski at Powder Mountain one last time; I suppose it was important for me to demonstrate that I was able to ski and function as before. But that year we put the condominiums up for sale, and I will never ski again. But the decision was probably Inger's and not mine.

Ivar windsurfing in Norway

Another sport that my whole family enjoyed immensely was windsurfing. I was introduced to the sport in 1976 at Lindau. One evening Carol Josephson, Brian's wife, had difficulty walking to dinner, so I asked her what the matter was. She told me she had taken a lesson in windsurfing and had hurt her leg. So I said: "What is windsurfing?" She explained that it is kind of like sailing a small boat, but you stood up on a board when you sailed. I had sailed a lot when I was young, but I had never heard of anything like this. She said it was kind of a surfboard with sails and it was invented in San Diego. This intrigued me because I was living in San Diego at the time, and I had not come across such a contraption. Since I had tried to surf when I was there, I became very interested and decided to investigate. She told me somebody was giving a course at a beach in Lindau. The next day I went there and sure enough somebody was giving a course on windsurfing. It was my last day in Lindau so I asked the guy if I could try it. He said of course, but you have to pay for the course, which was comprised of 10 lessons. I said I was leaving the next day and would like to try, but in typical German fashion, he said that in order to sail in Lake Constance you needed a license. I do not know if that was true or not, but we finally settled on $25 to try the board without any lessons. Since I had a little surfing experience from San Diego and was good at sailing, I managed the windsurfer right away and loved it. I sailed upwind for quite a distance with no problems and had some difficulty getting back, but finally made it.

As soon as I could I ordered a windsurfer from California and started to windsurf in local lakes. It became my favorite summer sport, and I introduced and instructed all my kids and Inger in the sport. Compared to skiing there are no lift lines and no crowds. You simply bring the windsurfer to the beach on top of your car, and it is easy to rig it up there. We soon had a couple of windsurfers both in Norway and at home; in particular I loved to windsurf in the Oslo fjord, because of all the islands outside my cottage in Norway.

CHAPTER 25

IN BUSINESS

We tried to grow the business by giving talks about the science at conferences and at the invitation of professors interested in the technology. This worked reasonably well. We also went to listen to talks and RPI had some business-focused talks referred to as "Varsity Talks". At one of these talks, Charlie listened to Chris Dehnert who gave a talk about how to sell scientific equipment. He listed 10 rules which described the wrong approach, and we were guilty of all of them! So we decided to meet with Chris, and after that there was no looking back. He joined our company as an adviser, and took no money. But our agreement was that he would be handsomely paid if he could make the company reach certain sales targets. He seemed very motivated and we got along well. It took longer than what he had anticipated in the beginning to hit those targets. When it was just Charlie and me, we often, talked about how we needed a catalogue but we never actually got around to designing one. Three weeks after Chris joined us, we had printed our first catalogue listing all our equipment for sale. Next he decided it was time to go to a trade show. When scientists host technical meetings, the business people or "vendors" go to the same meetings and pay for exhibit space to display their wares. It is an effective way to reach customers. Before we went we invested in a display tool and had large pictures taken of our equipment. I thought it was rather expensive, too fancy and a little over-the-top, but Charlie rationalized the cost by saying we could rent the equipment to other people in the incubator center. It was a reasonable argument, which of course we never followed through on.

There was an American Society for Cell Biology meeting in Orlando that we all went to and I found it very educational. Basically, you stand in front of your booth and try to catch people's interest. Charlie and I manned the booth and Chris was also there. He brought along an attractive woman who also worked in his private business. Chris clearly knew what he was doing; it always pays to have a beautiful woman to attract people to the booth. Once people stopped by, we could get their name and address to send them information and boast about how great our equipment is. I am actually rather shy, so when people approached us, I secretly hoped that they would not stop by the booth. Luckily Charlie is the opposite, he loves talking to people and waved them in for a chat. Since we were the first company to sell equipment that measured tissue culture cells electrically, we had quite a difficult time explaining the concept to potential customers. Now, more than 20 years later, we still get people at our booth who have never heard about our system. Some ideas take longer to incubate than others.

In addition to getting to know potential customers' names you also get to know what customers want. As scientists and inventors of the device we wanted to sell, we tend to think we know the best way to proceed with experiments. But this is not always the case, and it is hard to resist the temptation to lecture people, so not surprisingly,

Charlie and Ivar talking with potential customers

scientists are rarely the best salespeople. From this tradeshow we learned that that our small 16-well electrode arrays were too expensive for academic labs and that pharmaceutical customers wanted 96 wells. Chris taught us how to adapt.

We were a good team, and at the conferences we had dinner together. Being in Orlando, Florida there were many tourist things to try. We went to a vertical wind tunnel where you could fly without any equipment. We were 3 men and only one woman, but only the woman was brave enough to actually fly.

Our company operated as a small shop, with most of the manufacturing, quality control, and packaging all done by ourselves; we were a "start-up" before the term became fashionable. For example, when Nurayan who did the manufacturing of electrodes decided to leave, he was nice enough to recommend another RPI graduate student as a replacement, and we hired him. The new guy, Keith, was a good worker but he had strange ideas. Despite having a degree in engineering he was into mysticism, and he had a second degree in philosophy. He did a stint in India after graduation, and had married and divorced. He had a daughter he was devoted to. Because he held an engineering degree I assumed he would be rational, but I was wrong. One day he came to the office with a cardboard box. Inside this box he had placed some electrical equipment, that he claimed would cure cancer. When he left us after a few years, he ended up studying to become a chiropractor. Chris now decided we needed help and so we hired Jenn who took care of the office and was very efficient. Sales started to grow, and we started to strategize how best to grow the business.

We were just two scientists at Applied BioPhysics. Charlie took care of the cell cultures, and I took care of the instrument programming. I coded in APL, which stands for "A Programming Language" and originated at IBM. My daughter, Guri, who was studying engineering at Brown University, introduced me to APL. I asked her if she had learned any computer programming and she said: "Not really, I use APL, you write down the formulas you need and the computer gives you the answer." It turned out that my Norwegian friend Jens Feder was an expert in APL and convinced me to change from Basic to APL. I love the cryptic way APL is programmed, and there is a saying: "If God had known APL he could have created the world in one day!" I have no way of knowing why this happened but one day Iverson, the originator of APL at IBM, called me and said we are having a conference and can you come and give a talk?" To me I could not have been more surprised if God had called, so of course I accepted. APL notation was first constructed at Harvard by Iverson to teach people how to think logically. When Iverson joined IBM it was made into a computer program. I remember that in my talk I praised the program, but criticized its input and output features. I loved to program in APL, but I was weak at making actual graphs, and it was not easy to make our program user friendly.

Applied BioPhysics move in at the RPI Techpark

Applied Biophysics was growing and we needed more help, so we hired Jenn's husband Chris. We had quite a heated discussion before we hired him because it is in general not a good idea to have a married couple in a small firm. For example, they will take vacations together, and be together night and day. Chris worked as an independent contractor and although he did not have much knowledge of mathematics, he had an undergraduate degree in biology from Siena College. Both Chris and Jenn were very likeable and they became our good friends. Charlie used to say they could take over the business one day. They were hired as sales people and got a very good commission from Chris Dehnert. In my opinion, the commission was too high, but I was not very experienced in this regard.

Chris did not really understand the ECIS concept well. Furthermore, his sales expertise was not great. He was responsible for sales in Japan, but he could not answer their questions so he typically came to me for help. When we made a sale in Brazil, he had to go to Sao Paulo in

Applied Biophysics new location at RPI Techpark

Brazil to install the new machine. He was anxious, and did not do well. He went to China for us and did a better job demonstrating the machine. But since this was his first travel abroad, maybe it was to be expected. So despite these hiccups, we thought Chris would work out in the end.

At Applied BioPhysics we always have a Christmas party with nice meals and drinks. We give speeches about the business conditions and sometimes we give small bonuses. From the beginning, I was the CEO and Charlie was the CTO. At one Christmas I said in my speech that it was time to hire a *real* CEO. I had noticed that we shipped things every day, but recognized that I did not really know how it worked. Furthermore, I recognized that neither Charlie nor I were very good at business. This speech created a lot of commotion in the company. At this time we had two consultants: David a professional programmer and Don an electronics expert. Don was obviously was afraid he would lose his consultant agreement and came to talk to me. David was not worried and offered himself as a CEO, obviously as a joke. Jenn who tended to overestimate her ability wanted to be the new

CEO. Chris then got into an argument with Chris Dehnert because they felt entitled to the CEO job, but he was afraid that Dehnert would get it. While this upheaval was going on, sales had increased so Chris Dehnert earned his 25% ownership of the company. The happy family aspect of the company deteriorated. Both Jenn and Chris wrote us letters that criticized the way we had run the company; they each hoped to be CEO and in the end we had to fire them both. This was a sad ending of a beautiful friendship but business is business. Neither Chris Dehnert nor I saw any reasonable other solution but Charlie, who is kinder, was more hesitant.

ECIS travel

It is very difficult to do business in Japan for foreigners and Applied Biophysics was no exception. To serve the Japanese market, early on we hired a distributor in Japan, but he was not very effective. At one trade show a possible solution arose — we met a guy from Texas who talked up a Japanese company named Nepagene. He claimed he had connections and that they might work for us. He was going to a conference in Japan and suggested we accompany him. We discussed it internally and I was chosen to go, because a Nobel Prize carries a lot of weight in Japan. I was reluctant to go because I thought it was a bad investment. By this time I traveled only in business class, and I saw the trip as a $6000 gamble. Nepagene was going to pay for my stay in Japan. The guy from Texas had promised to meet me at the airport when I arrived, but of course he was not there. The day got worse from there. Without my "friend" from Texas, I had to find the train to downtown Tokyo. But the US had just rolled out brand new $100 bills, and no one in Japan would exchange them! Fortunately I had an old, wrinkled $50 bill which let me get to the downtown hotel. When I checked in, the guy who was supposed to meet me was staying there, so I went to bed thinking we would meet in the morning. When I called his room next morning he had already left for the conference. This pattern was becoming familiar, but still very annoying. To find him and the conference, I had to call back to the USA. Eventually I made my way to the conference. I was scheduled to give a talk right at lunchtime and arrived with very little time to spare. There were 3 speakers on the schedule, in a big hall holding 300–400 people. Everyone in the audience was eating a free boxed lunch that had been used to bribe them to the talks, so no one was really listening. The whole misadventure was a big disappointment to me. The next day I visited Nepagene which turned out to be a small firm much likes ours, and I thought it might be OK to go with them as our distributor. That night their management team took me out to dinner followed by drinks. The cocktails were served along with four or five very young and very beautiful girls who were working as lap dancers. My hosts had bought dance tickets and were very disappointed that I did not want to partake. I was 60 years old and told them I would need an older dancer. This comment, along with many others that evening, was lost in translation. The lesson I learned from this journey was that, Nobel Prize or not, it is unfortunately easy for a small business to waste $6,000. I am not sure if Chris and Charlie learned the same lesson?

ECIS meeting

In 1996 while I was still at RPI, I received a paper to review for the *Journal of Biochemical and Biophysical Methods*. The paper was by Wegener J, Sieber M and Galla H.J. and appeared to be an edited version of a Ph.D. thesis. The paper was basically a copy of our paper on ECIS, but to my surprise, they had not referenced our paper. I thought the paper was good and recommended publication, but I also pointed out that they should have referenced our paper. Shortly after reviewing that paper, I attended an impedance conference in Germany. There I met Joachim Wegener, one of the authors, who tried to initiate a conversation. I was rather rude because he had not referenced our paper, but a few years later Joachim wrote to me asking if he could come as a postdoc to RPI and that he had a fellowship to support his salary. I discussed it with Charlie, and he was not enthusiastic

First ECIS conference in Regensburg

about the idea, but I decided to let him come. That turned out to be a very good decision on my part. Joachim (or Joe as he likes to be called) was a very good worker and a very nice guy. I decided that his failure to cite our paper must have been his professor's fault. Since joining our team, we have published many papers together and he has made ECIS a big part of his career. Joe is now a professor in Regensburg, Germany, and in 2009 he arranged the first ECIS meeting. I did not think this was a good idea but to my amazement about 100 people showed up. We held our 2010 meeting locally, in Troy, NY. My contribution to the meeting was to give a talk entitled: "How to win the Nobel Prize." That meeting marked a new chapter for me. My talks are now more general and less technical, and they are intended to be fun. I have made a rule of not doing too many of these talks, but I still have a hard time turning down speaking requests from students, so I continue to rack up frequent flyer miles as I travel to India, Kazakhstan, South Korea and other interesting destinations.

Today

CHAPTER 26

There have been a lot of changes at Applied Biophysics in the past few years. While I originally did all the programming in APL, we wanted to make things more "user friendly" so I looked at other programming languages. After trying many options, I settled on the old standby known as "C". After getting pretty good at this language, I went to Norway for the summer, as Inger and I do every summer. On my return I discovered Charlie and Chris had hired David who used MATLAB, a language very similar to APL, so I did not object too strenuously and handed over the reins. On other fronts, we developed the 96-well plates that the Pharma industry wanted, and we hired Christian, a researcher with an undergraduate degree in physics and a graduate degree in biology. He was a great asset and had no difficulty understanding the workings of our instrument. And he likes to travel, so Christian also handles foreign sales.

I still come in to work every day, but to a large extent I have been phased out of day-to-day operations. I do not go to trade shows anymore and I have recently spent most of my time writing rather than running the business. At lunch I eat a cup of very spicy noodles that I buy at Walmart for 28 cents. Recently when I went there, I bought 3 packages with 12 cups in each. When I was going to pay, the lady at the cash registered said: "You don't have to pay. It is already paid for." I was puzzled by that but thanked her and went away. My friends explained to me later this is called "Paying It Forward" i.e. you pay for the person behind you in the line because you are kind and think he needs it. Whoever paid would have been shocked to know that he had paid for a Nobel laureate! In many ways this reminds me of an episode of the popular 1990's TV show, Seinfeld. I fell in love with that show from the beginning, and still laugh at the many situations that George, Jerry, Elaine, and Kramer encounter. At the moment I feel like a person in the episode where George offered to help Jerry buy a car. George is preoccupied with a chocolate bar he had lost in a vending machine, so he got upset and complained to the owner of the dealership. While he was describing the great chocolate bar injustice, George noticed a guy just sitting and watching the whole scene. So George asks: "Who is he?" The owner deadpans: "He is my dad. He is allowed to come in, but he is not allowed to talk to the customers."

That captures how I feel right now. I have had a good run, but the time has come to pull in the oars and simply enjoy the few years Inger and I have left.

Epilogue

In 1972, Philip Merilees wrote a paper called *"Does the flap of a butterfly's wings in Brazil set off a tornado in Texas?"* The weather forecasters have a big problem because as this title indicates, it is impossible to know all initial conditions. In chaos theory very small changes in the initial conditions sometimes have an enormous effect on the final results. This is also true in life; at least it is definitely true in my life. At various time in my life, I made trivial choices that turned out to have very enormous effects on what happened later. We simply call that luck. People in general have a large faith in luck that is why lotteries and casinos survive. I have been told that Las Vegas casinos sometimes fly big gamblers for free to their hotels and the gamblers take pride in this, but from the casino's point of view this is clearly a business decision. They obviously think you are a fool if you accept such an offer. I believe that people in general think that they are rational, but there are many cultural biases. For example, most gamblers think they are lucky. My daughters think that many men (probably including me) believe that they are smarter than women. I asked most of the Chinese students at RPI if they believed in acupuncture, and they all did. If your parents were very religious, chances are you also believe in God. Most people think that Nobel Prize winners are very smart. I once got interviewed by telephone from Norway, and the interviewer said: "You must be very smart" and I replied "Why do you think so?" "Because you have the Nobel Prize," he said. "No, no," I answered. "Some winners are smart, some are average and a few are actually dumb." The next day he wrote an article in the newspaper that said: "Professor Giaever says that Nobel Prize winners are stupid!"

I was recently at a dinner party and somebody asked: "Professor Giaever, what is the most important thing that has happened to you in your lifetime?" After some thought I said: "It has to be the internet. It is like living in a library." But then a woman across from the table said: "Don't you have any children?" So I had to change my statement, but it does not sound so good saying that the internet is the fifth most important thing in my life!

When Inger and I look back we find that we have been very fortunate. Our children have grown up to be responsible and nice adults with good education; John is an engineer, Anne is a teacher, Guri followed me and became a scientist, and Trine is an artist. When Trine wanted to go to Rhode Island School of Design, I told her she could, but she would not make any money. But after some time she got a job with *The Daily News* and made a lot of money. When I asked her why she worked there she said: When I paint only you and Mom look at my paintings, but in *The Daily News*, a million people look at my work every day!" Now, however, she is a mother and paints very successfully in her "spare" time. We are very proud of our kids; it cannot be too easy to have a famous father. We now have 8 grandchildren as well, but no great grandchildren yet to my wife's sorrow. She loves babies, so hopefully she will have her wish granted. I am more interested in technology and am amazed by how rapidly it has progressed. I am happy that the Nobel Prize Committee has started to recognize technology again; for example the blue LED light developed by three Japanese scientists in 2014. I worked at GE when the first red LED came on the market in 1962, and someone said they would be great for Christmas tree lights. At the time I did not recognize how fast technology develops and thought it was a crazy idea because LEDs were expensive. I recently bought a "60 watt" equivalent LED lightbulb for $10.90 and it is supposed to last for 22.2 years! Not a good buy if you are 86 years old! I firmly believe that the smartphone introduced by Apple deserves a Nobel Prize in physics. I recently participated in a discussion about technology and I said: "In the future we will never see so rapid a development of technology as we have seen in this century."

The whole family on vacation in 2008

Then somebody said: "What if we develop a computer that can really think?" Young people often ask me what they should work on in science, and I tell them if I were young I would work on artificial intelligence. If we can make a machine who can truly learn I am sure this would be the ultimate invention. I used to think that if a machine could play chess it must be smart, but the computers who can beat chess champions are merely cleverly programmed. However, if a computer could truly learn, simply by being told the rules, two computers could play each other and become chess champions.

Another question I also often get is: "How did it feel to get the Nobel Prize?" and now my answer is simply: "I suddenly became the most famous person I knew!"

www.ingramcontent.com/pod-product-compliance
Lightning Source LLC
Chambersburg PA
CBHW061935220426
43662CB00012B/1920